Biomedicine

Biomedicine

Edited by
Davon Matthews

Larsen & Keller
www.larsen-keller.com

Biomedicine
Edited by Davon Matthews
ISBN: 978-1-63549-047-3 (Hardback)

© 2017 Larsen & Keller

⊟ Larsen & Keller

Published by Larsen and Keller Education,
5 Penn Plaza,
19th Floor,
New York, NY 10001, USA

Cataloging-in-Publication Data

Biomedicine / edited by Davon Matthews.
        p. cm.
Includes bibliographical references and index.
ISBN 978-1-63549-047-3
1. Medicine. 2. Medical sciences. 3. Biology--Research.
I. Matthews, Davon.
R129 .B56 2017
610--dc23

The publisher's policy is to use permanent paper from mills that operate a sustainable forestry policy. Furthermore, the publisher ensures that the text paper and cover boards used have met acceptable environmental accreditation standards.

Printed and bound in the United States of America.

For more information regarding Larsen and Keller Education and its products, please visit the publisher's website www.larsen-keller.com

Table of Contents

Preface

This book elucidates the concepts and innovative models around prospective developments with respect to biomedicine. It talks in detail about the various branches that fall under this field. Biomedicine refers to the application of different principles of natural science and biological science to medical practice. It seeks to improve methods of diagnostic treatment as well as the various technologies of medical science. Most of the topics introduced in this text cover new techniques and the applications of biomedicine. Different approaches, evaluations and methodologies on the subject have been included in the book, thus making it an invaluable source of knowledge for the students.

To facilitate a deeper understanding of the contents of this book a short introduction of every chapter is written below:

Chapter 1- This chapter will provide an integrated understanding of biomedicine. Biomedicine is medicine based on the principles of natural sciences and biology and is applied to clinical practice. It seeks to provide effective interventional practice to various diseases and disorders. Biomedicine plays a huge role in people's understanding of heath and illness.

Chapter 2- A detailed account on gene therapy is explained in this chapter with regard to treating patients. Disciplines such as biochemistry, cell biology, virology etc. are also elaborated. Biomedicine is an emerging field and this chapter provides a profusion of interdisciplinary topics for better comprehension of biomedicine.

Chapter 3- The chapter serves as a source to understand the major technological breakthroughs of biomedicine. The major ideas dealt are health care, genome and genome project , human metabolome database etc. An informative account is given about health care and the prevention of disease, illness and injury. The themes discussed in the chapter are of great importance to broaden the existing knowledge on biomedicine.

Chapter 4- Medical diagnosis, differential diagnosis, therapy and the 2009 flu pandemic vaccine are the significant topics included in this chapter. The process of determining a person's illness is known as medical diagnosis, while differential diagnosis is differentiating one disease from another that presents similar clinical features. The topics discussed in the chapter are of great importance to broaden the existing knowledge on biomedicine.

Chapter 5- The illnesses elaborated in this chapter are HIV, HIV/AIDS and swine influenza. The chapter provides an integrated understanding on the current concerns surrounding biomedical research. Biomedicine has developed a reputation for focusing too much on treatment, rather than prevention, which cannot be very useful when it comes to illnesses like HIV and swine influenza. This chapter on HIV/AIDS and swine influenza offers an insightful focus, keeping in mind the complex subject matter.

Finally, I would like to thank the entire team involved in the inception of this book for their valuable time and contribution. This book would not have been possible without their efforts. I would also like to thank my friends and family for their constant support.

Editor

Introduction to Biomedicine

This chapter will provide an integrated understanding of biomedicine. Biomedicine is medicine based on the principles of natural sciences and biology and is applied to clinical practice. It seeks to provide effective interventional practice to various diseases and disorders. Biomedicine plays a huge role in people's understanding of heath and illness.

Biomedicine (i.e. *Medical biology*) is a branch of medical science that applies biological and other natural-science principles to clinical practice. The branch especially applies to biology and physiology. Biomedicine also can relate to many other categories in health and biological related fields. It has been the dominant health system for more than a century.

It includes many biomedical disciplines and areas of specialty that typically contain the "bio-" prefix such as:

- molecular biology, biochemistry, biotechnology, cell biology, embryology,
- nanobiotechnology, biological engineering, laboratory medical biology,
- cytogenetics, genetics, gene therapy,
- bioinformatics, biostatistics, systems biology, neuroscience,
- microbiology, virology, immunology, parasitology,
- physiology, pathology, anatomy,
- toxicology, and many others that generally concern life sciences as applied to medicine.

Medical biology is the cornerstone of modern health care and laboratory diagnostics. It concerns a wide range of scientific and technological approaches: from an in vitro diagnostics to the in vitro fertilisation, from the molecular mechanisms of a cystic fibrosis to the population dynamics of the HIV virus, from the understanding molecular interactions to the study of the carcinogenesis, from a single-nucleotide polymorphism (SNP) to the gene therapy.

Medical biology based on molecular biology combines all issues of developing molecular medicine into large-scale structural and functional relationships of the human genome, transcriptome, proteome, physiome and metabolome with the particular point of view of devising new technologies for prediction, diagnosis and therapy

Biomedicine involves the study of (patho-) physiological processes with methods from biology and physiology. Approaches range from understanding molecular interactions to the study of the consequences at the in vivo level. These processes are studied with the particular point of view of devising new strategies for diagnosis and therapy.

Depending on the severity of the disease, biomedicine pinpoints a problem within a patient and fixes the problem through medical intervention. Medicine focuses on curing diseases rather than improving one's health.

Molecular Biology

Molecular biology is the a process of synthesis and regulation of a cell's DNA, RNA, and protein. Molecular biology consists of different techniques including Polymerase chain reaction, Gel electrophoresis, and macromolecule blotting to manipulate DNA.

Polymerase chain reaction is done by placing a mixture of the desired DNA, DNA polymerase, primers, and nucleotide bases into a machine. The machine heats up and cools down at various temperatures to break the hydrogen bonds binding the DNA and allows the nucleotide bases to be added onto the two DNA templates after it has been separated.

Gel electrophoresis is a technique used to identify similar DNA between two unknown samples of DNA. This process is done by first preparing an agarose gel. This jelly-like sheet will have wells for DNA to be poured into. An electric current is applied so that the DNA, which is negatively charged due to its phosphate groups is attracted to the positive electrode. Different rows of DNA will move at different speeds because some DNA pieces are larger than others. Thus if two DNA samples show a similar pattern on the gel electrophoresis, one can tell that these DNA samples match.

Macromolecule blotting is a process performed after gel electrophoresis. An alkaline solution is prepared in a container. A sponge is placed into the solution and an agaros gel is placed on top of the sponge. Next, nitrocellulose paper is placed on top of the agarose gel and a paper towels are added on top of the nitrocellulose paper to apply pressure. The alkaline solution is drawn upwards towards the paper towel. During this process, the DNA denatures in the alkaline solution and is carried upwards to the nitrocellulose paper. The paper is then placed into a plastic bag and filled with a solution full of the DNA fragments, called the probe, found in the desired sample of DNA. The probes anneal to the complimentary DNA of the bands already found on the nitrocellulose sample. Afterwards, probes are washed off and the only ones present are the ones that have annealed to complimentary DNA on the paper. Next the paper is stuck onto a x ray film. The radioactivity of the probes creates black bands on the film, called an autoradiograph. As a result, only similar patterns of DNA to that of the probe are present on the film. This allows us the compare similar DNA sequences of multiple DNA samples. The overall process results in a precise reading of similarities in both similar and different DNA sample.

Biochemistry

Biochemistry is the science of the chemical processes which takes place within living organisms. Living organisms need essential elements to survive, consisting of carbon, hydrogen, nitrogen, oxygen, calcium, and phosphorus. These elements make up the four big macromolecules that living organisms need to survive- carbohydrates, lipids, proteins, and nucleic acids.

Carbohydrates, made up of carbon, hydrogen, and oxygen, are energy storing molecules. The simplest one of carbohydrates is glucose, $C_6H_{12}O_6$, is used in cellular respiration to produce ATP, adenosine triphosphate, which supplies cells with energy.

Proteins are chains of amino acids that function to contract skeletal muscle, function catalysts, transport molecules, and storage molecules. Proteins can facilitate biochemical processes, by lowering the activation energy of a reaction. Hemoglobins are also proteins, that carry oxygen to the cells in an organisms body.

Lipids, also known as fats, also serve to store energy, but in the long term. Due to their unique structure, lipids provide more than twice the amount of energy that carbohydrates do. Lipids can be used as insulation, as it is present below the layer of skin in living organisms. Moreover, lipids can be used in hormone production to maintain a healthy hormonal balance and provide structure to your cell walls.

Nucleic acids are used to store DNA in every living organism. The two types of nucleic acids are DNA and RNA. DNA is the main genetic information storing substance found oftentimes in the nucleus, which controls the processes that the cell undergoes. DNA consists of two complimentary antiparallel strands consisting varying patterns of nucleotides. RNA is a single strand of DNA, which is transcribed from DNA and used for DNA translation, which is the process for making proteins out of RNA sequences.

References

- "biomedicine (applies biological, physiological) - Memidex dictionary/thesaurus". memidex.com. 2012-10-08. Retrieved 2012-10-20.

- "University of Würzburg Graduate Schools: Biomedicine". graduateschools.uni-wuerzburg.de. 2011-10-14. Retrieved 2012-10-20.

2

Related Disciplines of Biomedicine

A detailed account on gene therapy is explained in this chapter with regard to treating patients. Disciplines such as biochemistry, cell biology, virology etc. are also elaborated. Biomedicine is an emerging field and this chapter provides a profusion of interdisciplinary topics for better comprehension of biomedicine.

Molecular Biology

Molecular biology concerns the molecular basis of biological activity between biomolecules in the various systems of a cell, including the interactions between DNA, RNA and proteins and their biosynthesis, as well as the regulation of these interactions. Writing in *Nature* in 1961, William Astbury described molecular biology as:

"...not so much a technique as an approach, an approach from the viewpoint of the so-called basic sciences with the leading idea of searching below the large-scale manifestations of classical biology for the corresponding molecular plan. It is concerned particularly with the forms of biological molecules and [...] is predominantly three-dimensional and structural—which does not mean, however, that it is merely a refinement of morphology. It must at the same time inquire into genesis and function."

Relationship to Other Biological Sciences

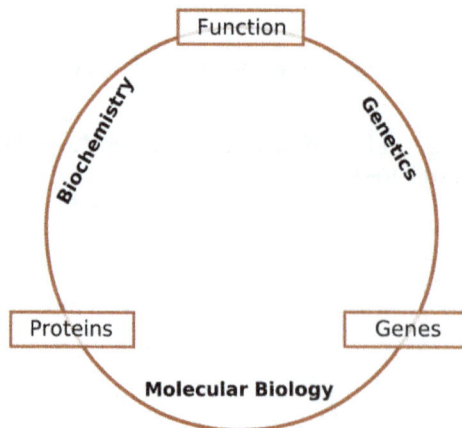

Schematic relationship between biochemistry, genetics and molecular biology

Researchers in molecular biology use specific techniques native to molecular biology but increasingly combine these with techniques and ideas from genetics and biochemistry. There is not a defined line between these disciplines. The figure to the right is a schematic that depicts one possible view of the relationship between the fields:

- *Biochemistry* is the study of the chemical substances and vital processes occurring in live organisms. Biochemists focus heavily on the role, function, and structure of biomolecules. The study of the chemistry behind biological processes and the synthesis of biologically active molecules are examples of biochemistry.

- *Genetics* is the study of the effect of genetic differences on organisms. This can often be inferred by the absence of a normal component (e.g. one gene). The study of "mutants" – organisms which lack one or more functional components with respect to the so-called "wild type" or normal phenotype. Genetic interactions (epistasis) can often confound simple interpretations of such "knockout" studies.

- *Molecular biology* is the study of molecular underpinnings of the processes of replication, transcription, translation, and cell function. The central dogma of molecular biology where genetic material is transcribed into RNA and then translated into protein, despite being an oversimplified picture of molecular biology, still provides a good starting point for understanding the field. This picture, however, is undergoing revision in light of emerging novel roles for RNA.

Much of the work in molecular biology is quantitative, and recently much work has been done at the interface of molecular biology and computer science in bioinformatics and computational biology. As of the early 2000s, the study of gene structure and function, molecular genetics, has been among the most prominent sub-field of molecular biology.Increasingly many other loops of biology focus on molecules, either directly studying their interactions in their own right such as in cell biology and developmental biology, or indirectly, where the techniques of molecular biology are used to infer historical attributes of populations or species, as in fields in evolutionary biology such as population genetics and phylogenetics. There is also a long tradition of studying biomolecules "from the ground up" in biophysics.

Techniques of Molecular Biology

Since the late 1950s and early 1960s, molecular biologists have learned to characterize, isolate, and manipulate the molecular components of cells and organisms. These components include DNA, the repository of genetic information; RNA, a close relative of DNA whose functions range from serving as a temporary working copy of DNA to actual structural and enzymatic functions as well as a functional and structural part of the translational apparatus, the ribosome; and proteins, the major structural and enzymatic type of molecule in cells.

Molecular Cloning

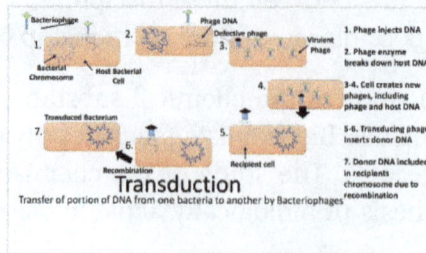

Transduction image

One of the most basic techniques of molecular biology to study protein function is molecular cloning. In this technique, DNA coding for a protein of interest is cloned (using PCR and/or restriction enzymes) into a plasmid (known as an expression vector). A vector has 3 distinctive features: an origin of replication, a multiple cloning site (MCS), and a selective marker (usually antibiotic resistance). The origin of replication will have promoter regions upstream from the replication/transcription start site.

This plasmid can be inserted into either bacterial or animal cells. Introducing DNA into bacterial cells can be done by transformation (via uptake of naked DNA), conjugation (via cell-cell contact) or by transduction (via viral vector). Introducing DNA into eukaryotic cells, such as animal cells, by physical or chemical means is called transfection. Several different transfection techniques are available, such as calcium phosphate transfection, electroporation, microinjection and liposome transfection. DNA can also be introduced into eukaryotic cells using viruses or bacteria as carriers, the latter is sometimes called bactofection and in particular uses *Agrobacterium tumefaciens*. The plasmid may be integrated into the genome, resulting in a stable transfection, or may remain independent of the genome, called transient transfection.

In either case, DNA coding for a protein of interest is now inside a cell, and the protein can now be expressed. A variety of systems, such as inducible promoters and specific cell-signaling factors, are available to help express the protein of interest at high levels. Large quantities of a protein can then be extracted from the bacterial or eukaryotic cell. The protein can be tested for enzymatic activity under a variety of situations, the protein may be crystallized so its tertiary structure can be studied, or, in the pharmaceutical industry, the activity of new drugs against the protein can be studied.

Polymerase Chain Reaction (PCR)

Polymerase chain reaction is an extremely versatile technique for copying DNA. In brief, PCR allows a specific DNA sequence to be copied or modified in predetermined ways. The reaction is extremely powerful and under perfect conditions could amplify 1 DNA molecule to become 1.07 Billion molecules in less than 2 hours. The PCR technique can be used to introduce restriction enzyme sites to ends of DNA molecules, or to mutate (change) particular bases of DNA, the latter is a method referred to as site-directed mutagenesis. PCR can also be used to determine whether a particular DNA fragment is found in a cDNA library. PCR has many variations, like reverse transcription PCR (RT-PCR) for amplification of RNA, and, more recently, quantitative PCR which allow for quantitative measurement of DNA or RNA molecules.

Gel Electrophoresis

Two percent Agarose Gel in Borate Buffer cast in a Gel Tray (Front, angled)

Gel electrophoresis is one of the principal tools of molecular biology. The basic principle is that DNA, RNA, and proteins can all be separated by means of an electric field and size. In agarose gel electrophoresis, DNA and RNA can be separated on the basis of size by running the DNA through an electrically charged agarose gel. Proteins can be separated on the basis of size by using an SDS-PAGE gel, or on the basis of size and their electric charge by using what is known as a 2D gel electrophoresis.

Macromolecule Blotting and Probing

The terms *northern*, *western* and *eastern* blotting are derived from what initially was a molecular biology joke that played on the term *Southern blotting*, after the technique described by Edwin Southern for the hybridisation of blotted DNA. Patricia Thomas, developer of the RNA blot which then became known as the *northern blot*, actually didn't use the term. Further combinations of these techniques produced such terms as *southwesterns* (protein-DNA hybridizations), *northwesterns* (to detect protein-RNA interactions) and *farwesterns* (protein-protein interactions), all of which are presently found in the literature.

Southern Blotting

Named after its inventor, biologist Edwin Southern, the Southern blot is a method for probing for the presence of a specific DNA sequence within a DNA sample. DNA samples before or after restriction enzyme (restriction endonuclease) digestion are separated by gel electrophoresis and then transferred to a membrane by blotting via capillary action. The membrane is then exposed to a labeled DNA probe that has a complement base sequence to the sequence on the DNA of interest. Most original protocols used radioactive labels, however non-radioactive alternatives are now available. Southern blotting is less commonly used in laboratory science due to the capacity of other techniques, such as PCR, to detect specific DNA sequences from DNA samples. These blots are still used for some applications, however, such as measuring transgene copy number in transgenic mice, or in the engineering of gene knockout embryonic stem cell lines.

Northern Blotting

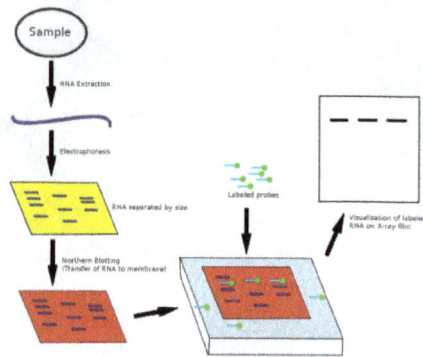

Northern blot diagram

The northern blot is used to study the expression patterns of a specific type of RNA molecule as relative comparison among a set of different samples of RNA. It is essentially a combination of denaturing RNA gel electrophoresis, and a blot. In this process RNA is separated based on size and is then transferred to a membrane that is then probed with a labeled complement of a sequence of interest. The results may be visualized through a variety of ways depending on the label used; however, most result in the revelation of bands representing the sizes of the RNA detected in sample. The intensity of these bands is related to the amount of the target RNA in the samples analyzed. The procedure is commonly used to study when and how much gene expression is occurring by measuring how much of that RNA is present in different samples. It is one of the most basic tools for determining at what time, and under what conditions, certain genes are expressed in living tissues.

Western Blotting

Antibodies to most proteins can be created by injecting small amounts of the protein into an animal such as a mouse, rabbit, sheep, or donkey (polyclonal antibodies) or produced in cell culture (monoclonal antibodies). These antibodies can be used for a variety of analytical and preparative techniques.

In western blotting, proteins are first separated by size, in a thin gel sandwiched between two glass plates in a technique known as SDS-PAGE (sodium dodecyl sulfate polyacrylamide gel electrophoresis). The proteins in the gel are then transferred to a polyvinylidene fluoride (PVDF), nitrocellulose, nylon, or other support membrane. This membrane can then be probed with solutions of antibodies. Antibodies that specifically bind to the protein of interest can then be visualized by a variety of techniques, including colored products, chemiluminescence, or autoradiography. Often, the antibodies are labeled with enzymes. When a chemiluminescent substrate is exposed to the enzyme it allows detection. Using western blotting techniques allows not only detection but also quantitative analysis. Analogous methods to western blotting can be used to directly stain specific proteins in live cells or tissue sections. However, these *immunostaining* methods, such as FISH, are used more often in cell biology research.

Eastern Blotting

The Eastern blotting technique is used to detect post-translational modification of proteins. Proteins blotted on to the PVDF or nitrocellulose membrane are probed for modifications using specific substrates.

Microarrays

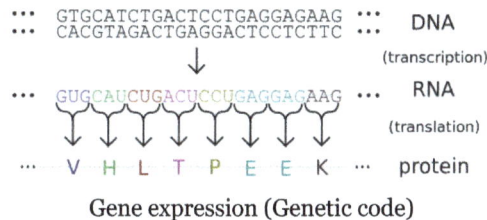

Gene expression (Genetic code)

A DNA microarray is a collection of spots attached to a solid support such as a microscope slide where each spot contains one or more single-stranded DNA oligonucleotide fragment. Arrays make it possible to put down large quantities of very small (100 micrometre diameter) spots on a single slide. Each spot has a DNA fragment molecule that is complementary to a single DNA sequence (similar to Southern blotting). A variation of this technique allows the gene expression of an organism at a particular stage in development to be qualified (expression profiling). In this technique the RNA in a tissue is isolated and converted to labeled cDNA. This cDNA is then hybridized to the fragments on the array and visualization of the hybridization can be done. Since multiple arrays can be made with exactly the same position of fragments they are particularly useful for comparing the gene expression of two different tissues, such as a healthy and cancerous tissue. Also, one can measure what genes are expressed and how that expression changes with time or with other factors. For instance, the common baker's yeast, *Saccharomyces cerevisiae*, contains about 7000 genes; with a microarray, one can measure qualitatively how each gene is expressed, and how that expression changes, for example, with a change in temperature. There are many different ways to fabricate microarrays; the most common are silicon chips, microscope slides with spots of ~ 100 micrometre diameter, custom arrays, and arrays with larger spots on porous membranes (macroarrays). There can be anywhere from 100 spots to more than 10,000 on a given array.Arrays can also be made with molecules other than DNA. For example, an antibody array can be used to determine what proteins or bacteria are present in a blood sample.

Allele-specific Oligonucleotide

Allele-specific oligonucleotide (ASO) is a technique that allows detection of single base mutations without the need for PCR or gel electrophoresis. Short (20-25 nucleotides in length), labeled probes are exposed to the non-fragmented target DNA. Hybridization occurs with high specificity due to the short length of the probes and even a single base change will hinder hybridization. The target DNA is then washed and the labeled probes that didn't hybridize are removed. The target DNA is then analyzed for the presence of the probe via radioactivity or fluorescence. In this experiment, as in most molecular biology techniques, a control must be used to ensure successful experimentation. The Illumina Methylation Assay is an example of a method that takes advantage of the ASO technique to measure one base pair differences in sequence.

Antiquated Technologies

In molecular biology, procedures and technologies are continually being developed and older technologies abandoned. For example, before the advent of DNA gel electrophoresis (agarose or polyacrylamide), the size of DNA molecules was typically determined by rate sedimentation in sucrose gradients, a slow and labor-intensive technique requiring expensive instrumentation; prior to sucrose gradients, viscometry was used. Aside from their historical interest, it is often worth knowing about older technology, as it is occasionally useful to solve another new problem for which the newer technique is inappropriate.

History

While molecular biology was established in the 1930s, the term was coined by Warren Weaver in 1938. Weaver was the director of Natural Sciences for the Rockefeller Foundation at the time and believed that biology was about to undergo a period of significant change given recent advances in fields such as X-ray crystallography. He therefore channeled significant amounts of (Rockefeller Institute) money into biological fields.

Clinical Significance

Clinical research and medical therapies arising from molecular biology are partly covered under gene therapy. The use of molecular biology or molecular cell biology approaches in medicine is now called molecular medicine. Molecular biology also plays important role in understanding formations, actions, regulations of various parts of cells which can be used efficiently for targeting new drugs, diagnosis of disease, physiology of the Cell.

Biochemistry

Biochemistry, sometimes called biological chemistry, is the study of chemical processes within and relating to living organisms. By controlling information flow through biochemical signaling and the flow of chemical energy through metabolism, biochemical processes give rise to the complexity of life. Over the last decades of the 20th century, biochemistry has become so successful at explaining living processes that now almost all areas of the life sciences from botany to medicine to genetics are engaged in biochemical research. Today, the main focus of pure biochemistry is on understanding how biological molecules give rise to the processes that occur within living cells, which in turn relates greatly to the study and understanding of tissues, organs, and whole organisms—that is, all of biology.

Biochemistry is closely related to molecular biology, the study of the molecular mechanisms by which genetic information encoded in DNA is able to result in the processes of life. Depending on the exact definition of the terms used, molecular biology can be thought of as a branch of biochemistry, or biochemistry as a tool with which to investigate and study molecular biology.

Much of biochemistry deals with the structures, functions and interactions of biological macromolecules, such as proteins, nucleic acids, carbohydrates and lipids, which provide the structure of cells and perform many of the functions associated with life. The chemistry of the cell also depends on the reactions of smaller molecules and ions. These can be inorganic, for example water and metal ions, or organic, for example the amino acids, which are used to synthesize proteins. The mechanisms by which cells harness energy from their environment via chemical reactions are known as metabolism. The findings of biochemistry are applied primarily in medicine, nutrition, and agriculture. In medicine, biochemists investigate the causes and cures of diseases. In nutrition, they study how to maintain health and study the effects of nutritional deficiencies. In agriculture, biochemists investigate soil and fertilizers, and try to discover ways to improve crop cultivation, crop storage and pest control.

History

At its broadest definition, biochemistry can be seen as a study of the components and composition of living things and how they come together to become life, and the history of biochemistry may therefore go back as far as the ancient Greeks. However, biochemistry as a specific scientific discipline has its beginning some time in the 19th century, or a little earlier, depending on which aspect of biochemistry one is being focused on. Some argued that the beginning of biochemistry may have been the discovery of the first enzyme, diastase (today called amylase), in 1833 by Anselme Payen, while others considered Eduard Buchner's first demonstration of a complex biochemical process alcoholic fermentation in cell-free extracts in 1897 to be the birth of biochemistry. Some might also point as its beginning to the influential 1842 work by Justus von Liebig, *Animal chemistry, or, Organic chemistry in its applications to physiology and pathology*, which presented a chemical theory of metabolism, or even earlier to the 18th century studies on fermentation and respiration by Antoine Lavoisier. Many other pioneers in the field who helped to uncover the layers of complexity of biochemistry have been proclaimed founders of modern biochemistry, for example Emil Fischer for his work on the chemistry of proteins, and F. Gowland Hopkins on enzymes and the dynamic nature of biochemistry.

Gerty Cori and Carl Cori jointly won the Nobel Prize in 1947 for their discovery of the Cori cycle at RPMI.

The term "biochemistry" itself is derived from a combination of biology and chemistry. In 1877, Felix Hoppe-Seyler used the term (*biochemie* in German) as a synonym for physiological chemistry in the foreword to the first issue of *Zeitschrift für Physiologische Chemie* where he argued for the setting up of institutes dedicated to this field of study. The German chemist Carl Neuberg however is often cited to have been coined the word in 1903, while some credited it to Franz Hofmeister.

DNA structure (1D65)

It was once generally believed that life and its materials had some essential property or substance (often referred to as the "vital principle") distinct from any found in non-living matter, and it was thought that only living beings could produce the molecules of life. Then, in 1828, Friedrich Wöhler published a paper on the synthesis of urea, proving that organic compounds can be created artificially. Since then, biochemistry has advanced, especially since the mid-20th century, with the development of new techniques such as chromatography, X-ray diffraction, dual polarisation interferometry, NMR spectroscopy, radioisotopic labeling, electron microscopy, and molecular dynamics simulations. These techniques allowed for the discovery and detailed analysis of many molecules and metabolic pathways of the cell, such as glycolysis and the Krebs cycle (citric acid cycle).

Another significant historic event in biochemistry is the discovery of the gene and its role in the transfer of information in the cell. This part of biochemistry is often called molecular biology. In the 1950s, James D. Watson, Francis Crick, Rosalind Franklin, and Maurice Wilkins were instrumental in solving DNA structure and suggesting its relationship with genetic transfer of information. In 1958, George Beadle and Edward Tatum received the Nobel Prize for work in fungi showing that one gene produces one enzyme. In 1988, Colin Pitchfork was the first person convicted of murder with DNA evidence, which led to growth of forensic science. More recently, Andrew Z. Fire and Craig C. Mello received the 2006 Nobel Prize for discovering the role of RNA interference (RNAi), in the silencing of gene expression.

Starting Materials: The Chemical Elements of Life

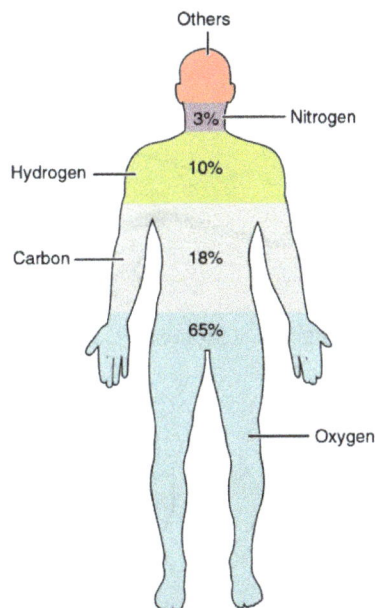

The main elements that compose the human body are shown from most abundant (by mass) to least abundant.

Around two dozen of the 92 naturally occurring chemical elements are essential to various kinds of biological life. Most rare elements on Earth are not needed by life (exceptions being selenium and iodine), while a few common ones (aluminum and titanium) are not used. Most organisms share element needs, but there are a few differences between plants and animals. For example, ocean algae use bromine, but land plants and animals seem to need none. All animals require sodium, but some plants do not. Plants need boron and silicon, but animals may not (or may need ultra-small amounts).

Just six elements—carbon, hydrogen, nitrogen, oxygen, calcium, and phosphorus—make up almost 99% of the mass of living cells, including those in the human body. In addition to the six major elements that compose most of the human body, humans require smaller amounts of possibly 18 more.

Biomolecules

The four main classes of molecules in biochemistry (often called biomolecules) are carbohydrates, lipids, proteins, and nucleic acids. Many biological molecules are polymers: in this terminology, *monomers* are relatively small micromolecules that are linked together to create large macromolecules known as *polymers*. When monomers are linked together to synthesize a biological polymer, they undergo a process called dehydration synthesis. Different macromolecules can assemble in larger complexes, often needed for biological activity.

Carbohydrates

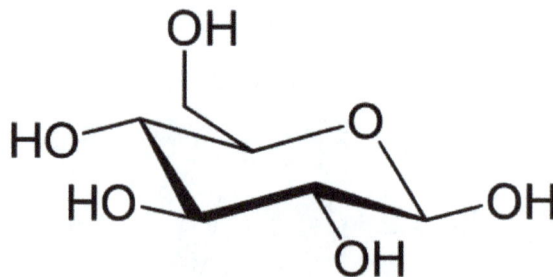

Glucose, a monosaccharide

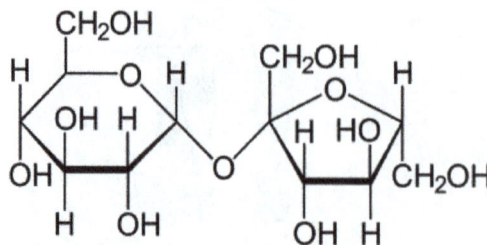

A molecule of sucrose (glucose + fructose), a disaccharide

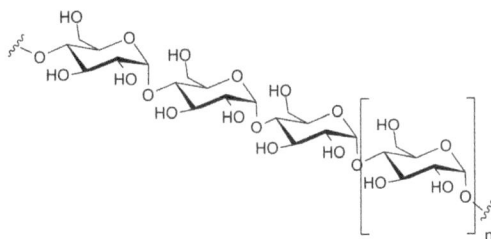

Amylose, a polysaccharide made up of several thousand glucose units

The function of carbohydrates includes energy storage and providing structure. Sugars are carbohydrates, but not all carbohydrates are sugars. There are more carbohydrates on Earth than any other known type of biomolecule; they are used to store energy and genetic information, as well as play important roles in cell to cell interactions and communications.

The simplest type of carbohydrate is a monosaccharide, which among other properties contains carbon, hydrogen, and oxygen, mostly in a ratio of 1:2:1 (generalized formula $C_nH_{2n}O_n$, where n is at least 3). Glucose ($C_6H_{12}O_6$) is one of the most important carbohydrates, others include fructose ($C_6H_{12}O_6$), the sugar commonly associated with the sweet taste of fruits, and deoxyribose ($C_5H_{10}O_4$).

A monosaccharide can switch from the acyclic (open-chain) form to a cyclic form, through a nucleophilic addition reaction between the carbonyl group and one of the hydroxyls of the same molecule. The reaction creates a ring of carbon atoms closed by one bridging oxygen atom. The resulting molecule has an hemiacetal or hemiketal group, depending on whether the linear form was an aldose or a ketose. The reaction is easily reversed, yielding the original open-chain form.

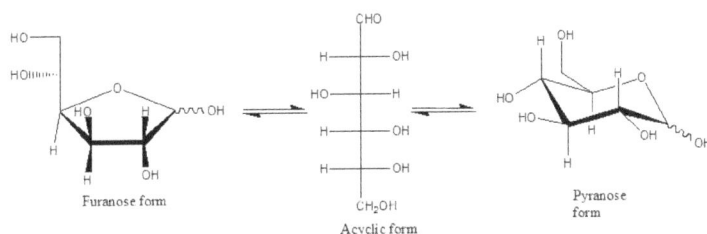

Conversion between the furanose, acyclic, and pyranose forms of D-glucose.

In these cyclic forms, the ring usually has 5 or 6 atoms. These forms are called furanoses and pyranoses, respectively — by analogy with furan and pyran, the simplest compounds with the same carbon-oxygen ring (although they lack the double bonds of these two molecules). For example, the aldohexose glucose may form a hemiacetal linkage between the hydroxyl on carbon 1 and the oxygen on carbon 4, yielding a molecule with a 5-membered ring, called glucofuranose. The same reaction can take place between carbons 1 and 5 to form a molecule with a 6-membered ring, called glucopyranose. Cyclic forms with a 7-atom ring (the same of oxepane), rarely encountered, are called heptoses.

When two monosaccharides undergo dehydration synthesis whereby a molecule of water is released, as two hydrogen atoms and one oxygen atom are lost from the two monosaccharides. The new molecule, consisting of two monosaccharides, is called a *disaccharide* and is conjoined together by a glycosidic or ether bond. The reverse reaction can also occur, using a molecule of water to split up a disaccharide and break the glycosidic bond; this is termed *hydrolysis*. The most well-known disaccharide is sucrose, ordinary sugar (in scientific contexts, called *table sugar* or *cane sugar* to differentiate it from other sugars). Sucrose consists of a glucose molecule and a fructose molecule joined together. Another important disaccharide is lactose, consisting of a glucose molecule and a galactose molecule. As most humans age, the production of lactase, the enzyme that hydrolyzes lactose back into glucose and galactose, typically decreases. This results in lactase deficiency, also called *lactose intolerance*.

When a few (around three to six) monosaccharides are joined, it is called an *oligosaccharide* (*oligo-* meaning "few"). These molecules tend to be used as markers and signals, as well as having some other uses. Many monosaccharides joined together make a polysaccharide. They can be joined together in one long linear chain, or they may be branched. Two of the most common polysaccharides are cellulose and glycogen, both consisting of repeating glucose monomers. Examples are *Cellulose* which is an important structural component of plant's cell walls, and *glycogen*, used as a form of energy storage in animals.

Sugar can be characterized by having reducing or non-reducing ends. A reducing end of a carbohydrate is a carbon atom that can be in equilibrium with the open-chain aldehyde (aldose) or keto form (ketose). If the joining of monomers takes place at such a carbon atom, the free hydroxy group of the pyranose or furanose form is exchanged with an OH-side-chain of another sugar, yielding a full acetal. This prevents opening of the chain to the aldehyde or keto form and renders the modified residue non-reducing. Lactose contains a reducing end at its glucose moiety, whereas the galactose moiety form a full acetal with the C4-OH group of glucose. Saccharose does not have a reducing end because of full acetal formation between the aldehyde carbon of glucose (C1) and the keto carbon of fructose (C2).

Lipids

Lipids comprises a diverse range of molecules and to some extent is a catchall for relatively water-insoluble or nonpolar compounds of biological origin, including waxes, fatty acids, fatty-acid derived phospholipids, sphingolipids, glycolipids, and terpenoids (e.g., retinoids and steroids). Some lipids are linear aliphatic molecules, while others have ring structures. Some are aromatic, while others are not. Some are flexible, while others are rigid.

Lipids are usually made from one molecule of glycerol combined with other molecules. In triglycerides, the main group of bulk lipids, there is one molecule of glycerol and three fatty acids. Fatty acids are considered the monomer in that case, and may be saturated (no double bonds in the carbon chain) or unsaturated (one or more double bonds in the carbon chain).

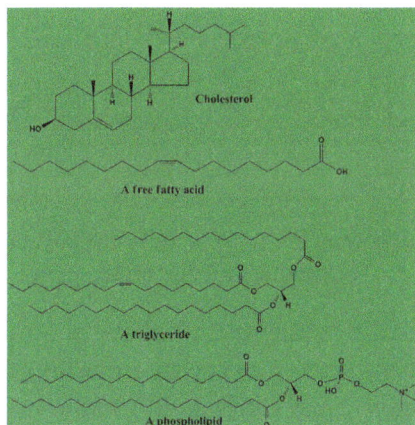

Structures of some common lipids. At the top are cholesterol and oleic acid. The middle structure is a triglyceride composed of oleoyl, stearoyl, and palmitoyl chains attached to a glycerol backbone. At the bottom is the common phospholipid, phosphatidylcholine.

Most lipids have some polar character in addition to being largely nonpolar. In general, the bulk of their structure is nonpolar or hydrophobic ("water-fearing"), meaning that it does not interact well with polar solvents like water. Another part of their structure is polar or hydrophilic ("water-loving") and will tend to associate with polar solvents like water. This makes them amphiphilic molecules (having both hydrophobic and hydrophilic portions). In the case of cholesterol, the polar group is a mere -OH (hydroxyl or alcohol). In the case of phospholipids, the polar groups are considerably larger and more polar, as described below.

Lipids are an integral part of our daily diet. Most oils and milk products that we use for cooking and eating like butter, cheese, ghee etc., are composed of fats. Vegetable oils are rich in various polyunsaturated fatty acids (PUFA). Lipid-containing foods undergo digestion within the body and are broken into fatty acids and glycerol, which are the final degradation products of fats and lipids. Lipids, especially phospholipids, are also used in various pharmaceutical products, either as co-solubilisers (e.g., in parenteral infusions) or else as drug carrier components (e.g., in a liposome or transfersome).

Proteins

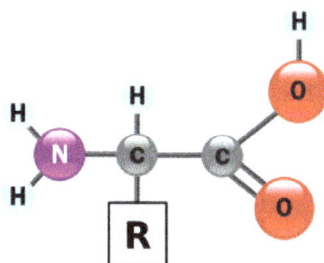

The general structure of an α-amino acid, with the amino group on the left and the carboxyl group on the right.

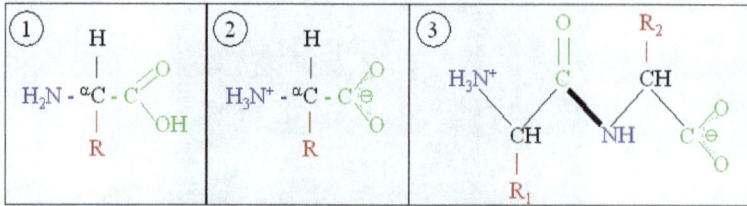

Generic amino acids (1) in neutral form, (2) as they exist physiologically, and (3) joined together as a dipeptide.

Proteins are very large molecules – macro-biopolymers – made from monomers called amino acids. An amino acid consists of a carbon atom bound to four groups. One is an amino group, $-NH_2$, and one is a carboxylic acid group, $-COOH$ (although these exist as $-NH_3^+$ and $-COO^-$ under physiologic conditions). The third is a simple hydrogen atom. The fourth is commonly denoted "$-R$" and is different for each amino acid. There are 20 standard amino acids, each containing a carboxyl group, an amino group, and a side-chain (known as an "R" group). The "R" group is what makes each amino acid different, and the properties of the side-chains greatly influence the overall three-dimensional conformation of a protein. Some amino acids have functions by themselves or in a modified form; for instance, glutamate functions as an important neurotransmitter. Amino acids can be joined via a peptide bond. In this dehydration synthesis, a water molecule is removed and the peptide bond connects the nitrogen of one amino acid's amino group to the carbon of the other's carboxylic acid group. The resulting molecule is called a *dipeptide*, and short stretches of amino acids (usually, fewer than thirty) are called *peptides* or polypeptides. Longer stretches merit the title *proteins*. As an example, the important blood serum protein albumin contains 585 amino acid residues.

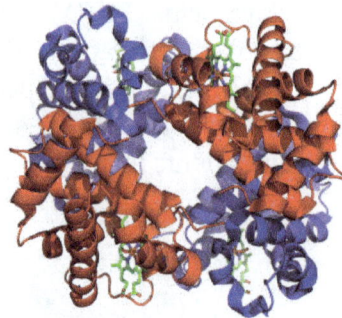

A schematic of hemoglobin. The red and blue ribbons represent the protein globin; the green structures are the heme groups.

Some proteins perform largely structural roles. For instance, movements of the proteins actin and myosin ultimately are responsible for the contraction of skeletal muscle. One property many proteins have is that they specifically bind to a certain molecule or class of molecules—they may be *extremely* selective in what they bind. Antibodies are an example of proteins that attach to one specific type of molecule. In fact, the enzyme-linked immunosorbent assay (ELISA), which uses antibodies, is one of the most sensitive tests modern medicine uses to detect various biomolecules. Probably the most important proteins, however, are the enzymes. Virtually every reaction in a living cell requires an enzyme to lower the activation energy of the reaction. These molecules recognize specific reactant molecules called *substrates*; they then catalyze the reaction between them. By lowering the activation energy, the enzyme speeds up that reaction by a rate of 10^{11} or more; a reaction that would normally take over 3,000 years to complete spontaneously might take less than a second with an enzyme. The enzyme itself is not used up in the process, and is free to catalyze the same reaction with a new set of substrates. Using various modifiers, the activity of the enzyme can be regulated, enabling control of the biochemistry of the cell as a whole.

The structure of proteins is traditionally described in a hierarchy of four levels. The primary structure of a protein simply consists of its linear sequence of amino acids; for instance, "alanine-glycine-tryptophan-serine-glutamate-asparagine-glycine-lysine-...". Secondary structure is concerned with local morphology (morphology being the study of structure). Some combinations of amino acids will tend to curl up in a coil called an α-helix or into a sheet called a β-sheet; some α-helixes can be seen in the hemoglobin schematic above. Tertiary structure is the entire three-dimensional shape of the protein. This shape is determined by the sequence of amino acids. In fact, a single change can change the entire structure. The alpha chain of hemoglobin contains 146 amino acid residues; substitution of the glutamate residue at position 6 with a valine residue changes the behavior of hemoglobin so much that it results in sickle-cell disease. Finally, quaternary structure is concerned with the structure of a protein with multiple peptide subunits, like hemoglobin with its four subunits. Not all proteins have more than one subunit.

Examples of protein structures from the Protein Data Bank

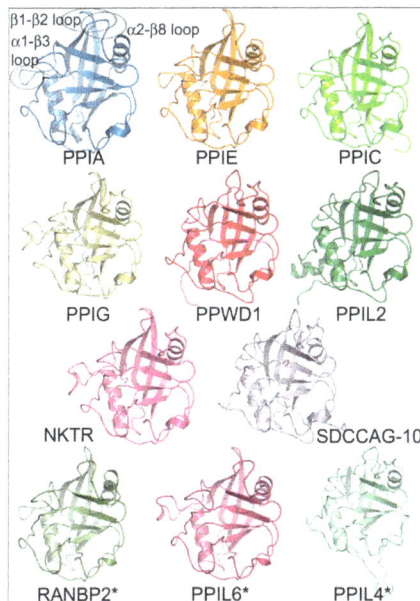

Members of a protein family, as represented by the structures of the isomerase domains.

Ingested proteins are usually broken up into single amino acids or dipeptides in the small intestine, and then absorbed. They can then be joined to make new proteins. Intermediate products of glycolysis, the citric acid cycle, and the pentose phosphate pathway can be used to make all twenty amino acids, and most bacteria and plants possess all the necessary enzymes to synthesize them. Humans and other mammals, however, can synthesize only half of them. They cannot synthesize isoleucine, leucine, lysine, methionine, phenylalanine, threonine, tryptophan, and valine. These are the essential amino acids, since it is essential to ingest them. Mammals do possess the enzymes to synthesize alanine, asparagine, aspartate, cysteine, glutamate, glutamine, glycine, proline, serine, and tyrosine, the nonessential amino acids. While they can synthesize arginine and histidine, they cannot produce it in sufficient amounts for young, growing animals, and so these are often considered essential amino acids.

If the amino group is removed from an amino acid, it leaves behind a carbon skeleton called an α-keto acid. Enzymes called transaminases can easily transfer the amino group from one amino acid (making it an α-keto acid) to another α-keto acid (making it an amino acid). This is important in the biosynthesis of amino acids, as for many of the pathways, intermediates from other biochemical pathways are converted to the α-keto acid skeleton, and then an amino group is added, often via transamination. The amino acids may then be linked together to make a protein.

A similar process is used to break down proteins. It is first hydrolyzed into its component amino acids. Free ammonia (NH_3), existing as the ammonium ion (NH_4^+) in blood, is toxic to life forms. A suitable method for excreting it must therefore exist. Different tactics have evolved in different animals, depending on the animals' needs. Unicellular organisms, of course, simply release the ammonia into the environment. Likewise, bony fish can release the ammonia into the water where it is quickly diluted. In general, mammals convert the ammonia into urea, via the urea cycle.

In order to determine whether two proteins are related, or in other words to decide whether they are homologous or not, scientists use sequence-comparison methods. Methods like sequence alignments and structural alignments are powerful tools that help scientists identify homologies between related molecules. The relevance of finding homologies among proteins goes beyond forming an evolutionary pattern of protein families. By finding how similar two protein sequences are, we acquire knowledge about their structure and therefore their function.

Nucleic Acids

Nucleic acids, so called because of its prevalence in cellular nuclei, is the generic name of the family of biopolymers. They are complex, high-molecular-weight biochemical macromolecules that can convey genetic information in all living cells and viruses. The monomers are called nucleotides, and each consists of three components: a nitrogenous heterocyclic base (either a purine or a pyrimidine), a pentose sugar, and a phosphate group.

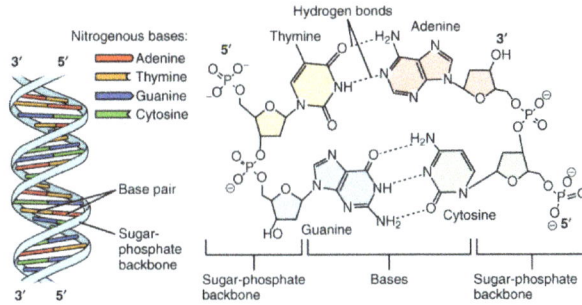

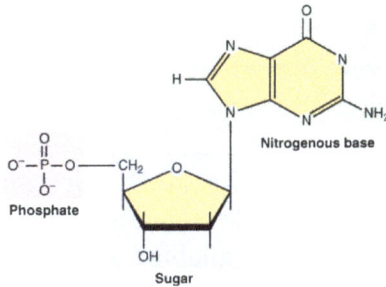

The structure of deoxyribonucleic acid (DNA), the picture shows the monomers being put together.

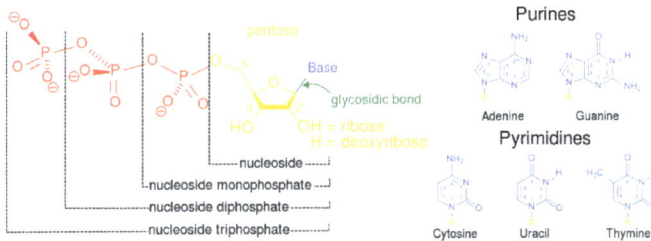

Structural elements of common nucleic acid constituents. Because they contain at least one phosphate group, the compounds marked *nucleoside monophosphate, nucleoside diphosphate* and *nucleoside triphosphate* are all nucleotides (not simply phosphate-lacking nucleosides).

The most common nucleic acids are deoxyribonucleic acid (DNA) and ribonucleic acid (RNA). The phosphate group and the sugar of each nucleotide bond with each other to form the backbone of the nucleic acid, while the sequence of nitrogenous bases stores the information. The most common nitrogenous bases are adenine, cytosine, guanine, thymine, and uracil. The nitrogenous bases of each strand of a nucleic acid will form hydrogen bonds with certain other nitrogenous bases in a complementary strand of nucleic acid (similar to a zipper). Adenine binds with thymine and uracil; Thymine binds only with adenine; and cytosine and guanine can bind only with one another.

Aside from the genetic material of the cell, nucleic acids often play a role as second messengers, as well as forming the base molecule for adenosine triphosphate (ATP), the primary energy-carrier molecule found in all living organisms. Also, the nitrogenous bases possible in the two nucleic acids are different: adenine, cytosine, and guanine occur in both RNA and DNA, while thymine occurs only in DNA and uracil occurs in RNA.

Metabolism

Carbohydrates as Energy Source

Glucose is the major energy source in most life forms. For instance, polysaccharides are broken down into their monomers (glycogen phosphorylase removes glucose residues from glycogen). Disaccharides like lactose or sucrose are cleaved into their two component monosaccharides.

Glycolysis (Anaerobic)

Glucose is mainly metabolized by a very important ten-step pathway called glycolysis, the net result of which is to break down one molecule of glucose into two molecules of pyruvate. This also produces a net two molecules of ATP, the energy currency of cells, along with two reducing equivalents of converting NAD^+ (nicotinamide adenine dinucleotide:oxidised form) to NADH (nicotinamide adenine dinucleotide:reduced form). This does not require oxygen; if no oxygen is available (or the cell cannot use oxygen), the NAD is restored by converting the pyruvate to lactate (lactic acid) (e.g., in humans) or to ethanol plus carbon dioxide (e.g., in yeast). Other monosaccharides like galactose and fructose can be converted into intermediates of the glycolytic pathway.

Aerobic

In aerobic cells with sufficient oxygen, as in most human cells, the pyruvate is further metabolized. It is irreversibly converted to acetyl-CoA, giving off one carbon atom as the waste product carbon dioxide, generating another reducing equivalent as NADH. The two molecules acetyl-CoA (from one molecule of glucose) then enter the citric acid cycle, producing two more molecules of ATP, six more NADH molecules and two reduced (ubi)quinones (via $FADH_2$ as enzyme-bound cofactor), and releasing the remaining carbon atoms as carbon dioxide. The produced NADH and quinol molecules then feed into the enzyme complexes of the respiratory chain, an electron transport system transferring the electrons ultimately to oxygen and conserving the released energy in the form of a proton gradient over a membrane (inner mitochondrial membrane in eukaryotes). Thus, oxygen is reduced to water and the original electron acceptors NAD^+ and quinone are regenerated. This is why humans breathe in oxygen and breathe out carbon dioxide. The energy released from transferring the electrons from high-energy states in NADH and quinol is conserved first as proton gradient and converted to ATP via ATP synthase. This generates an additional *28* molecules of ATP (24 from the 8 NADH + 4 from the 2 quinols), totaling to 32 molecules of ATP conserved per degraded glucose (two from glycolysis + two from the citrate cycle). It is clear that using oxygen to completely oxidize glucose provides an organism with far more energy than any oxygen-independent metabolic feature, and this is thought to be the reason why complex life appeared only after Earth's atmosphere accumulated large amounts of oxygen.

Gluconeogenesis

In vertebrates, vigorously contracting skeletal muscles (during weightlifting or sprinting, for example) do not receive enough oxygen to meet the energy demand, and so they shift to anaerobic metabolism, converting glucose to lactate. The liver regenerates the glucose, using a process called gluconeogenesis. This process is not quite the opposite of glycolysis, and actually requires three times the amount of energy gained from glycolysis (six molecules of ATP are used, compared to the two gained in glycolysis). Analogous to the above reactions, the glucose produced can then undergo glycolysis in tissues that need energy, be stored as glycogen (or starch in plants), or be converted to other monosaccharides or joined into di- or oligosaccharides. The combined pathways of glycolysis during exercise, lactate's crossing via the bloodstream to the liver, subsequent gluconeogenesis and release of glucose into the bloodstream is called the Cori cycle.

Relationship to Other "Molecular-scale" Biological Sciences

Researchers in biochemistry use specific techniques native to biochemistry, but increasingly combine these with techniques and ideas developed in the fields of genetics, molecular biology and biophysics. There has never been a hard-line among these disciplines in terms of content and technique. Today, the terms *molecular biology* and *biochemistry* are nearly interchangeable. The following figure is a schematic that depicts one possible view of the relationship between the fields:

- *Biochemistry* is the study of the chemical substances and vital processes occurring in living organisms. Biochemists focus heavily on the role, function, and structure of biomolecules. The study of the chemistry behind biological processes and the synthesis of biologically active molecules are examples of biochemistry.

- *Genetics* is the study of the effect of genetic differences on organisms. Often this can be inferred by the absence of a normal component (e.g., one gene). The study of "mutants" – organisms with a changed gene that leads to the organism being different with respect to the so-called "wild type" or normal phenotype. Genetic interactions (epistasis) can often confound simple interpretations of such "knock-out" or "knock-in" studies.

- *Molecular biology* is the study of molecular underpinnings of the process of replication, transcription and translation of the genetic material. The central dogma of molecular biology where genetic material is transcribed into RNA and then translated into protein, despite being an oversimplified picture of molecular biology, still provides a good starting point for understanding the field. This picture, however, is undergoing revision in light of emerging novel roles for RNA.

- *Chemical biology* seeks to develop new tools based on small molecules that allow minimal perturbation of biological systems while providing detailed information about their function. Further, chemical biology employs biological systems to create non-natural hybrids between biomolecules and synthetic devices (for example emptied viral capsids that can deliver gene therapy or drug molecules).

Cell Biology

Cell Structure

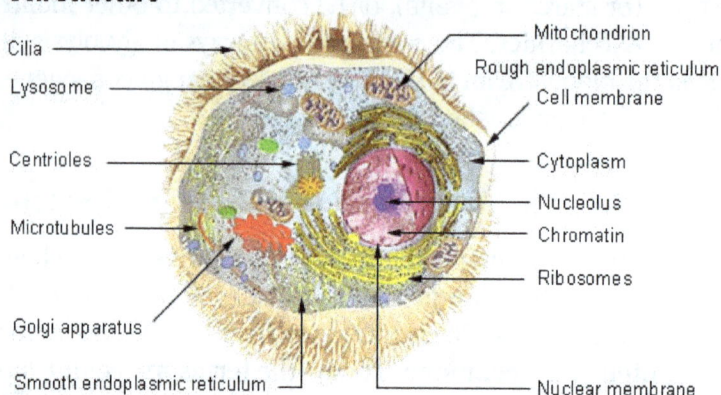

Understanding the cell in terms of its generalized structure and molecular components.

Cell biology and otherwise known as molecular biology, is a branch of biology that studies the different structures and functions of the cell and focuses mainly on the idea of the cell as the basic unit of life. Cell biology explains the structure, organization of the organelles they contain, their physiological properties, metabolic processes, signaling pathways, life cycle, and interactions with their environment. This is done both on a microscopic and molecular level as it encompasses prokaryotic cells and eukaryotic cells. Knowing the components of cells and how cells work is fundamental to all biological sciences it is also essential for research in bio-medical fields such as cancer, and other diseases. Research in cell biology is closely related to genetics, biochemistry, molecular biology, immunology, and developmental biology.

Internal Cellular Structures

Chemical and Molecular Environment

The study of the cell is done on a molecular level; however, most of the processes within the cell is made up of a mixture of small organic molecules, inorganic ions, hormones, and water. Approximately 75-85% of the cell's volume is due to water making it an indispensable solvent as a result of its polarity and structure. These molecules within the cell, which operate as substrates, provide a suitable environment for the cell to carry out metabolic reactions and signalling. The cell shape varies among the different types of organisms, and are thus then classified into two categories: eukaryotes and prokaryotes. In the case of eukaryotic cells - which are made up of animal, plant, fungi, and protozoa cells - the shapes are generally round and spherical, while for prokaryotic cells – which are composed of bacteria and archaea - the shapes are: spherical (cocci), rods (bacillus), curved (vibrio), and spirals (*spirochetes*).

Cell biology focuses more on the study of eukaryotic cells, and their signalling pathways, rather than on prokaryotes which is covered under microbiology. The main constituents of the general molecular composition of the cell includes: proteins and lipids which are either free flowing or membrane bound, along with different internal compartments known as organelles. This environment of the cell is made up of hydrophilic and hydrophobic regions which allows for the exchange of the above-mentioned molecules and ions. The hydrophilic regions of the cell are mainly on the inside and outside of the cell, while the hydrophobic regions are within the phospholipid bilayer of the cell membrane. The cell membrane consists of lipids and proteins which accounts for its hydrophobicity as a result of being non-polar substances. Therefore, in order for these molecules to participate in reactions, within the cell, they need to be able to cross this membrane layer to get into the cell. They accomplish this process of gaining access to the cell via: osmotic pressure, diffusion, concentration gradients, and membrane channels. Inside of the cell are extensive internal sub-cellular membrane-bounded compartments called organelles.

Organelles

- Centrosome - an associated pair of cylindrical shaped protein structures (centrioles) that organize microtubules and aid in forming the mitotic spindle during cell division in eukaryotes

- Cell membrane (plasma membrane) - the part of the cell which separates the cells from the outside environment and protects the cell, as well as regulating what goes in and out of the cell

- Cell wall - extra layer of protection and gives structural support (only found in plant cells)

- Chloroplast - key organelle for photosynthesis (only found in plant cells)

- Cilium - motile structure of eukaryotes having a cytoskeleton, the axoneme.

- Cytoplasm - contents of the main fluid-filled space inside cells, chemical reactions also happen in this jelly-like substance.

- Cytoskeleton - protein filaments inside cells (microfilaments, microtubules, and intermediate filaments)

- Endoplasmic reticulum (rough) - major site of membrane protein synthesis

- Endoplasmic reticulum (smooth) - major site of lipid synthesis

- Endosomes - vesicles that traffic membrane and intra and extra cellular contents for recycling or degradation by lysosomes

- Flagellum - motile structure of bacteria, archaea and eukaryotes

- Golgi apparatus - site of protein glycosylation in the endomembrane system

- Lipid bilayer - fundamental organizational structure of cell membranes

- Lysosome - acidic organelle that breaks down cellular waste products and debris into simple compounds (only found in animal cells)

- Microvilli - increases surface area for absorption of nutrients from surrounding medium

- Mitochondrion - major energy-producing organelle by releasing energy in the form of ATP

- Nucleus - contains chromosomes composed of DNA, the building block of life. Nuclear Architecture is important for dictating nuclear function.

- Organelle - term used for major subcellular structures

- Peroxisomes - a very small organelle that uses oxygen to breakdown and detoxify long fatty acids and other molecules

- Pili - also called fimbria is used for conjugation and sometimes movement

- Ribosome - RNA and protein complex required for protein synthesis in cells

- Starch grain - found in the cytoplasm of a typical plant cell, it stores chemical energy of the plant.

- Vacuole - contain cell sap (only found in plant cells)

- Vesicle - small membrane-bounded spheres inside cells

cell surface membrane protects the cell

Processes

Growth and Development

The growth process of the cell does not refer to the size of the cell, but instead the density of the number of cells present in the organism at a given time. Cell growth pertains to the increase in the number of cells present in an organism as it grows and develops; as the organism gets larger so too does the number of cells present.

Cells are the foundation of all organisms, they are the fundamental unit of life. The growth and development of the cell are essential for the maintenance of the host, and survival of the organisms. For this process the cell goes through the steps of the cell cycle and development which involves cell growth, DNA replication, cell division, regeneration, specialization, and cell death. The cell cycle is divided into four distinct phases, G1, S, G2, and M. The G phases – which is the cell growth phase - makes up approximately 95% of the cycle. The proliferation of cells is instigated by progenitors, the cells then differentiate to become specialized, where specialized cells of the same type aggregate to form tissues, then organs and ultimately systems. The G phases along with the S phase – DNA replication, damage and repair - are considered to be the interphase portion of the cycle. While the M phase (mitosis and cytokinesis) is the cell division portion of the cycle. The cell cycle is regulated by a series of signalling factors and complexes such as CDK's, kinases, and p53. to name a few. When the cell has completed its growth process, and if it is found to be damaged or altered it under-goes cell death, either by apoptosis or necrosis, to eliminate the threat it cause to the organism's survival.

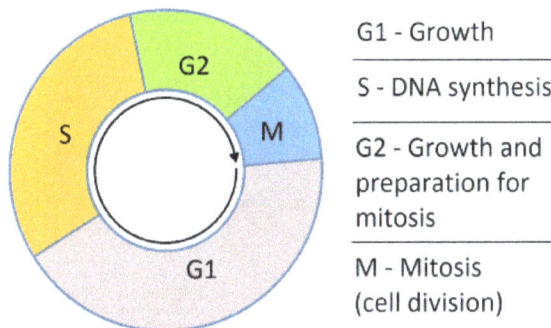

G1 - Growth

S - DNA synthesis

G2 - Growth and preparation for mitosis

M - Mitosis (cell division)

General concept of the cell cycle.

Other Cellular Processes

- Active transport and Passive transport - Movement of molecules into and out of cells.

- Autophagy - The process whereby cells "eat" their own internal components or microbial invaders.

- Adhesion - Holding together cells and tissues.

- Cell movement - Chemotaxis, contraction, cilia and flagella.

- Cell signaling - Regulation of cell behavior by signals from outside.

- Division - By which cells reproduce either by mitosis (to produce clones of the parent cell) or Meiosis (to produce haploid gametes)

- DNA repair - Cell death and cell senescence.

- Metabolism - Glycolysis, respiration, photosynthesis, and chemosynthesis.

- Signalling - The process by which the activities in the cell are regulated

- Transcription and mRNA splicing - Gene expression.

Techniques Used to Study Cells

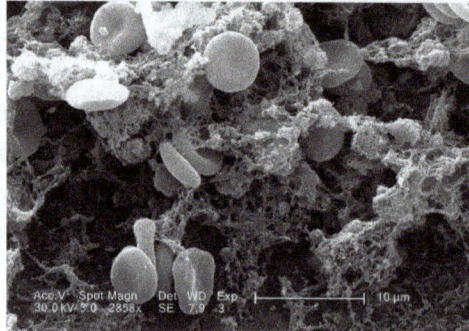

Electron micrograph of blood cells clotting.

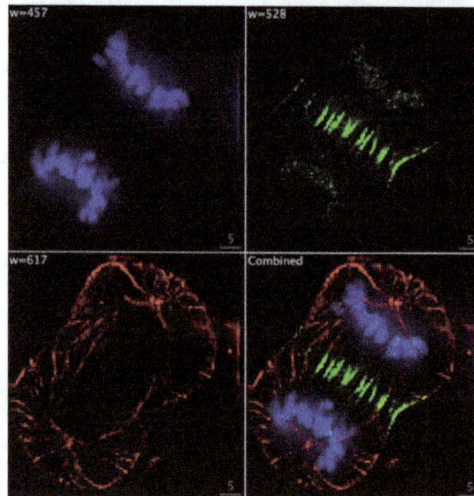

Cell division studied using fluorescence to stain specific structures

Cells may be observed under the microscope, using several different techniques; these include optical microscopy, transmission electron microscopy, scanning electron microscopy, fluorescence microscopy, and confocal microscopy.

There are several different methods used in the study of cells:

- Cell culture is the basic technique of growing cells in a laboratory independent of an organism.

- Immunostaining, also known as immunohistochemistry, is a specialized histological method used to localize proteins in cells or tissue slices. Unlike regular histology, which uses stains to identify cells, cellular components or protein classes, immunostaining requires the reaction of an antibody directed against the protein of interest within the tissue or cell. Through the use of proper controls and published protocols (need to add reference links here), specificity of the antibody-antigen reaction can be achieved. Once this complex is formed, it is identified via either a "tag" attached directly to the antibody, or added in an additional technical step. Commonly used "tags" include fluorophores or enzymes. In the case of the former, detection of the location of the "immuno-stained" protein occurs via fluorescence microscopy. With an enzymatic tag, such as horse radish peroxidase, a chemical reaction is carried out that results in a dark color in the location of the protein of interest. This darkened pattern is then detected using light microscopy.

- Computational genomics is used to find patterns in genomic information

- DNA microarrays identify changes in transcript levels between different experimental conditions.

- Gene knockdown mutates a selected gene.

- In situ hybridization shows which cells are expressing a particular RNA transcript.

- PCR can be used to determine how many copies of a gene are present in a cell.

- Transfection introduces a new gene into a cell, usually an expression construct

Purification of cells and their parts Purification may be performed using the following methods:

- Cell fractionation

  - Release of cellular organelles by disruption of cells.

  - Separation of different organelles by centrifugation.

- Flow cytometry

- Immunoprecipitation

  - The binding of an antibody to a target protein

  - Collection of the target protein through elution

- Proteins extracted from cell membranes by detergents and salts or other kinds of chemicals.

Embryology

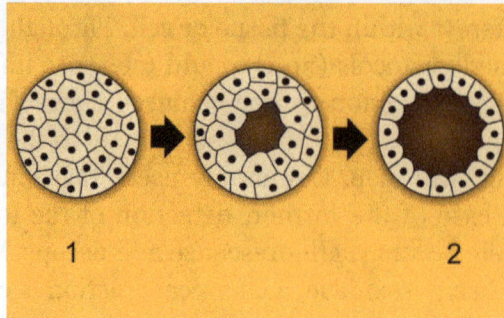

1 - morula, 2 - blastula

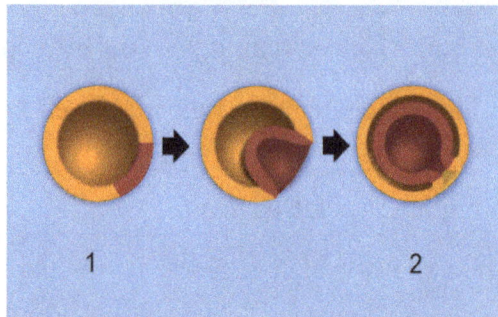

1 - blastula, 2 - gastrula with blastopore; orange - ectoderm, red - endoderm.

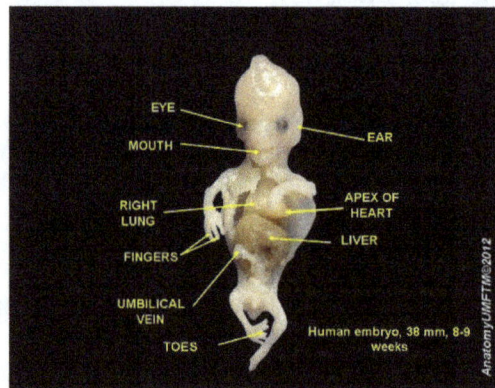

Dissection of man embryo, 38 mm - 8 weeks

Embryology is the branch of biology that studies the development of gametes (sex cells), fertilization, and development of embryos and fetuses. Additionally, embryology is the study of congenital disorders that occur before birth.

Embryonic Development of Animals

After cleavage, the dividing cells, or morula, becomes a hollow ball, or blastula, which develops a hole or pore at one end.

Bilaterals

In bilateral animals, the blastula develops in one of two ways that divides the whole animal kingdom into two halves. If in the blastula the first pore (blastopore) becomes the mouth of the animal, it is a protostome; if the first pore becomes the anus then it is a deuterostome. The protostomes include most invertebrate animals, such as insects, worms and molluscs, while the deuterostomes include the vertebrates. In due course, the blastula changes into a more differentiated structure called the gastrula.

The gastrula with its blastopore soon develops three distinct layers of cells (the germ layers) from which all the bodily organs and tissues then develop:

- The innermost layer, or endoderm, gives a rise to the digestive organs, the gills, lungs or swim bladder if present, and kidneys or nephrites.

- The middle layer, or mesoderm, gives rise to the muscles, skeleton if any, and blood system.

- The outer layer of cells, or ectoderm, gives rise to the nervous system, including the brain, and skin or carapace and hair, bristles, or scales.

Embryos in many species often appear similar to one another in early developmental stages. The reason for this similarity is because species have a shared evolutionary history. These similarities among species are called homologous structures, which are structures that have the same or similar function and mechanism, having evolved from a common ancestor.

Drosophila Melanogaster (Fruity Fly)

Drosophila melanogaster, a fruit fly, is a model organism in biology on which much research into embryology has been done. Before fertilization, the female gamete produces an abundance of mRNA - transcribed from the genes that encode bicoid protein and nanos protein. These mRNA molecules are stored to be used later in what will become a developing embryo. The male and female *Drosophila* gametes exhibit anisogamy (differences in morphology and sub-cellular biochemistry). The female gamete is larger than the male gamete because it harbors more cytoplasm and, within the cytoplasm, the female gamete contains an abundance of the mRNA previously mentioned. At fertilization, the male and female gametes fuse (plasmogamy) and then the nucleus of the male gamete fuses with the nucleus of the female gamete (karyogamy). Note that before the gametes' nuclei fuse, they are known as pronuclei. A series of nuclear divisions will occur without cytokinesis (division of the cell) in the zygote to form a multi-nucleated cell (a cell containing multiple nuclei) known as a syncytium. All the nuclei in the syncytium are identical, just as all the nuclei in every somatic cell of any multicellular organism are identical *in terms of the DNA sequence of the genome*. Before the nuclei can differentiate in transcriptional activity, the embryo (syncytium) must be divided into segments. In each segment, a unique set of regulatory proteins will cause specific genes in the nuclei to be transcribed. The resulting combination of proteins will transform clusters of cells into early embryo tissues that will each develop into multiple fetal and adult tissues later in development (note: this happens after each nucleus becomes wrapped with its own cell membrane).

Figure. A Drosophila melanogaster

Figure. B Drosophila melanogaster larvae contained in lab apparatus to be used for experiments in genetics and embryology.

Outlined below is the process that leads to cell and tissue differentiation.

Maternal-effect genes - subject to Maternal (cytoplasmic) inheritance.

- Egg-polarity genes establish the Anteroposterior axis.

Zygotic-effect genes - subject to Mendelian (classical) inheritance.

- Segmentation genes establish 14 segments of the embryo using the anteroposterior axis as a guide.

1. Gap genes establish 3 broad segments of the embryo.

2. Pair-rule genes define 7 segments of the embryo within the confines of the second broad segment that was defined by the gap genes.

3. Segment-polarity genes define another 7 segments by dividing each of the pre-existing 7 segments into anterior and posterior halves.

*   Homeotic (homeobox) genes use the 14 segments as pinpoints for specific types of cell differentiation and the histological developments that correspond to each cell type.

Humans

In humans, the term embryo refers to the ball of dividing cells from the moment the zygote implants itself in the uterus wall until the end of the eighth week after conception. Beyond the eighth week after conception (tenth week of pregnancy), the developing human is then called a fetus.

History

Human embryo at six weeks gestational age

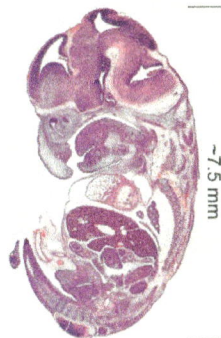

Histological film 10-day mouse embryo

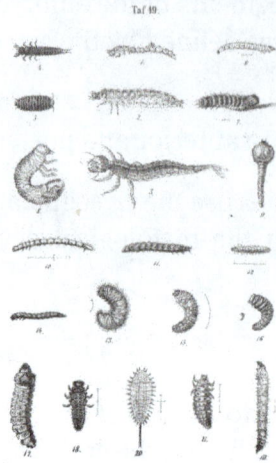

Beetle larvae

As recently as the 18th century, the prevailing notion in western human embryology was preformation: the idea that semen contains an embryo – a preformed, miniature infant, or *homunculus* – that simply becomes larger during development. The competing explanation of embryonic development was *epigenesis*, originally proposed 2,000 years earlier by Aristotle. Much early embryology came from the work of the Italian anatomists Aldrovandi, Aranzio, Leonardo da Vinci, Marcello Malpighi, Gabriele Falloppio, Girolamo Cardano, Emilio Parisano, Fortunio Liceti, Stefano Lorenzini, Spallanzani, Enrico Sertoli, and Mauro Rusconi. According to epigenesis, the form of an animal emerges gradually from a relatively formless egg. As microscopy improved during the 19th century, biologists could see that embryos took shape in a series of progressive steps, and epigenesis displaced preformation as the favoured explanation among embryologists.

After 1827

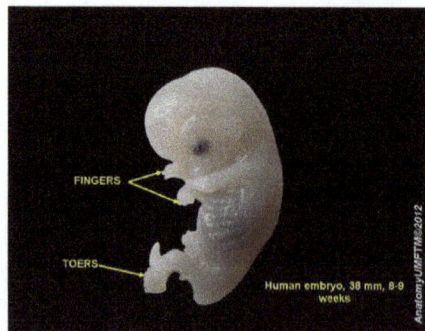

8–9-week human embryo

Karl Ernst von Baer and Heinz Christian Pander proposed the germ layer theory of development; von Baer discovered the mammalian ovum in 1827. Modern embryological pioneers include Charles Darwin, Ernst Haeckel, J.B.S. Haldane, and Joseph Needham. Other important contributors include William Harvey, Kaspar Friedrich Wolff, Heinz Christian Pander, August Weismann, Gavin de Beer, Ernest Everett Just, and Edward B. Lewis.

After 1950

After the 1950s, with the DNA helical structure being unravelled and the increasing knowledge in the field of molecular biology, developmental biology emerged as a field of study which attempts to correlate the genes with morphological change, and so tries to determine which genes are responsible for each morphological change that takes place in an embryo, and how these genes are regulated.

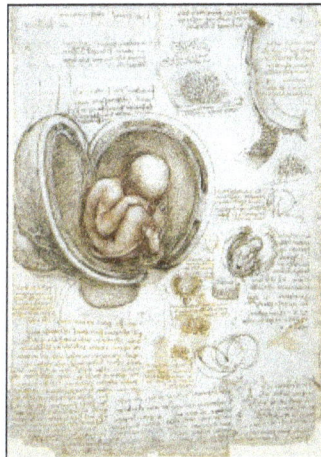

A study of embryos by Leonardo da Vinci

Vertebrate and Invertebrate Embryology

Many principles of embryology apply to invertebrates as well as to vertebrates. Therefore, the study of invertebrate embryology has advanced the study of vertebrate embryology. However, there are many differences as well. For example, numerous invertebrate species release a larva before development is complete; at the end of the larval period, an animal for the first time comes to resemble an adult similar to its parent or parents. Although invertebrate embryology is similar in some ways for different invertebrate animals, there are also countless variations. For instance, while spiders proceed directly from egg to adult form, many insects develop through at least one larval stage.

Modern Embryology Research

Currently, embryology has become an important research area for studying the genetic control of the development process (e.g. morphogens), its link to cell signalling, its importance for the study of certain diseases and mutations, and in links to stem cell research.

Nanobiotechnology

Nanobiotechnology, bionanotechnology, and nanobiology are terms that refer to the intersection of nanotechnology and biology. Given that the subject is one that has only emerged very recently, bionanotechnology and nanobiotechnology serve as blanket terms for various related technologies.

This discipline helps to indicate the merger of biological research with various fields of nanotechnology. Concepts that are enhanced through nanobiology include: nanodevices (such as biological machines), nanoparticles, and nanoscale phenomena that occurs within the discipline of nanotechnology. This technical approach to biology allows scientists to imagine and create systems that can be used for biological research. Biologically inspired nanotechnology uses biological systems as the inspirations for technologies not yet created. However, as with nanotechnology and biotechnology, bionanotechnology does have many potential ethical issues associated with it.

The most important objectives that are frequently found in nanobiology involve applying nanotools to relevant medical/biological problems and refining these applications. Developing new tools, such as peptoid nanosheets, for medical and biological purposes is another primary objective in nanotechnology. New nanotools are often made by refining the applications of the nanotools that are already being used. The imaging of native biomolecules, biological membranes, and tissues is also a major topic for the nanobiology researchers. Other topics concerning nanobiology include the use of cantilever array sensors and the application of nanophotonics for manipulating molecular processes in living cells.

Recently, the use of microorganisms to synthesize functional nanoparticles has been of great interest. Microorganisms can change the oxidation state of metals. These microbial processes have opened up new opportunities for us to explore novel applications, for example, the biosynthesis of metal nanomaterials. In contrast to chemical and physical methods, microbial processes for synthesizing nanomaterials can be achieved in aqueous phase under gentle and environmentally benign conditions. This approach has become an attractive focus in current green bionanotechnology research towards sustainable development.

Terminology

The terms are often used interchangeably. When a distinction is intended, though, it is based on whether the focus is on applying biological ideas or on studying biology with nanotechnology. Bionanotechnology generally refers to the study of how the goals of nanotechnology can be guided by studying how biological "machines" work and adapting these biological motifs into improving existing nanotechnologies or creating new ones. Nanobiotechnology, on the other hand, refers to the ways that nanotechnology is used to create devices to study biological systems.

In other words, nanobiotechnology is essentially miniaturized biotechnology, whereas bionanotechnology is a specific application of nanotechnology. For example, DNA nanotechnology or cellular engineering would be classified as bionanotechnology because they involve working with biomolecules on the nanoscale. Conversely, many new medical technologies involving nanoparticles as delivery systems or as sensors would be examples of nanobiotechnology since they involve using nanotechnology to advance the goals of biology.

The definitions enumerated above will be utilized whenever a distinction between nanobio and bionano is made in this article. However, given the overlapping usage of the terms in modern parlance, individual technologies may need to be evaluated to determine which term is more fitting. As such, they are best discussed in parallel.

Concepts

Most of the scientific concepts in bionanotechnology are derived from other fields. Biochemical principles that are used to understand the material properties of biological systems are central in bionanotechnology because those same principles are to be used to create new technologies. Material properties and applications studied in bionanoscience include mechanical properties(e.g. deformation, adhesion, failure), electrical/electronic (e.g. electromechanical stimulation, capacitors, energy storage/batteries), optical (e.g. absorption, luminescence, photochemistry), thermal (e.g. thermomutability, thermal management), biological (e.g. how cells interact with nanomaterials, molecular flaws/defects, biosensing, biological mechanisms s.a. mechanosensing), nanoscience of disease (e.g. genetic disease, cancer, organ/tissue failure), as well as computing (e.g. DNA computing). The impact of bionanoscience, achieved through structural and mechanistic analyses of biological processes at nanoscale, is their translation into synthetic and technological applications through nanotechnology.

Nano-biotechnology takes most of its fundamentals from nanotechnology. Most of the devices designed for nano-biotechnological use are directly based on other existing nanotechnologies. Nano-biotechnology is often used to describe the overlapping multidisciplinary activities associated with biosensors, particularly where photonics, chemistry, biology, biophysics, nano-medicine, and engineering converge. Measurement in biology using wave guide techniques, such as dual polarization interferometry, are another example.

Applications

Applications of bionanotechnology are extremely widespread. Insofar as the distinction holds, nanobiotechnology is much more commonplace in that it simply provides more tools for the study of biology. Bionanotechnology, on the other hand, promises to recreate biological mechanisms and pathways in a form that is useful in other ways.

Nanomedicine

Nanomedicine is a field of medical science whose applications are increasing more and more thanks to nanorobots and biological machines, which constitute a very useful tool to develop this area of knowledge. In the past years, researchers have done many improvements in the different devices and systems required to develop nanorobots. This supposes a new way of treating and dealing with diseases such as cancer; thanks to nanorobots, side effects of chemotherapy have been controlled, reduced and even eliminated, so some years from now, cancer patients will be offered an alternative to treat this disease instead of chemotherapy, which causes secondary effects such as hair loss, fatigue or nausea killing not only cancerous cells but also the healthy ones. At a clinical level, cancer treatment with nanomedicine will consist on the supply of nanorobots to the patient through an injection that will seek for cancerous cells leaving untouched the healthy ones. Patients that will be treated through nanomedicine will not notice the presence of this nanomachines inside them; the only thing that is going to be noticeable is the progressive improvement of their health.

Nanobiotechnology

Nanobiotechnology (sometimes referred to as nanobiology) is best described as helping modern medicine progress from treating symptoms to generating cures and regenerating biological tissues. Three American patients have received whole cultured bladders with the help of doctors who use nanobiology techniques in their practice. Also, it has been demonstrated in animal studies that a uterus can be grown outside the body and then placed in the body in order to produce a baby. Stem cell treatments have been used to fix diseases that are found in the human heart and are in clinical trials in the United States. There is also funding for research into allowing people to have new limbs without having to resort to prosthesis. Artificial proteins might also become available to manufacture without the need for harsh chemicals and expensive machines. It has even been surmised that by the year 2055, computers may be made out of biochemicals and organic salts.

Another example of current nanobiotechnological research involves nanospheres coated with fluorescent polymers. Researchers are seeking to design polymers whose fluorescence is quenched when they encounter specific molecules. Different polymers would detect different metabolites. The polymer-coated spheres could become part of new biological assays, and the technology might someday lead to particles which could be introduced into the human body to track down metabolites associated with tumors and other health problems. Another example, from a different perspective, would be evaluation and therapy at the nanoscopic level, i.e. the treatment of Nanobacteria (25-200 nm sized) as is done by NanoBiotech Pharma.

While nanobiology is in its infancy, there are a lot of promising methods that will rely on nanobiology in the future. Biological systems are inherently nano in scale; nanoscience must merge with biology in order to deliver biomacromolecules and molecular machines that are similar to nature. Controlling and mimicking the devices and processes that are constructed from molecules is a tremendous challenge to face the converging disciplines of nanotechnology. All living things, including humans, can be considered to be nanofoundries. Natural evolution has optimized the "natural" form of nanobiology over millions of years. In the 21st century, humans have developed the technology to artificially tap into nanobiology. This process is best described as "organic merging with synthetic." Colonies of live neurons can live together on a biochip device; according to research from Dr. Gunther Gross at the University of North Texas. Self-assembling nanotubes have the ability to be used as a structural system. They would be composed together with rhodopsins; which would facilitate the optical computing process and help with the storage of biological materials. DNA (as the software for all living things) can be used as a structural proteomic system - a logical component for molecular computing. Ned Seeman - a researcher at New York University - along with other researchers are currently researching concepts that are similar to each other.

Bionanotechnology

DNA nanotechnology is one important example of bionanotechnology. The utilization of the inherent properties of nucleic acids like DNA to create useful materials is a promising area of modern research. Another important area of research involves taking advantage of membrane properties to generate synthetic membranes. Proteins that self-assemble to generate functional materials could be used as a novel approach for the large-scale production of programmable nanomaterials. One example is the development of amyloids found in bacterial biofilms as engineered nanomaterials that can be programmed genetically to have different properties. Protein folding studies provide a third important avenue of research, but one that has been largely inhibited by our inability to predict protein folding with a sufficiently high degree of accuracy. Given the myriad uses that biological systems have for proteins, though, research into understanding protein folding is of high importance and could prove fruitful for bionanotechnology in the future.

Lipid nanotechnology is another major area of research in bionanotechnology, where physico-chemical properties of lipids such as their antifouling and self-assembly is exploited to build nanodevices with applications in medicine and engineering.

Tools

This field relies on a variety of research methods, including experimental tools (e.g. imaging, characterization via AFM/optical tweezers etc.), x-ray diffraction based tools, synthesis via self-assembly, characterization of self-assembly (using e.g. MP-SPR, DPI, recombinant DNA methods, etc.), theory (e.g. statistical mechanics, nanomechanics, etc.), as well as computational approaches (bottom-up multi-scale simulation, supercomputing).

Gene Therapy

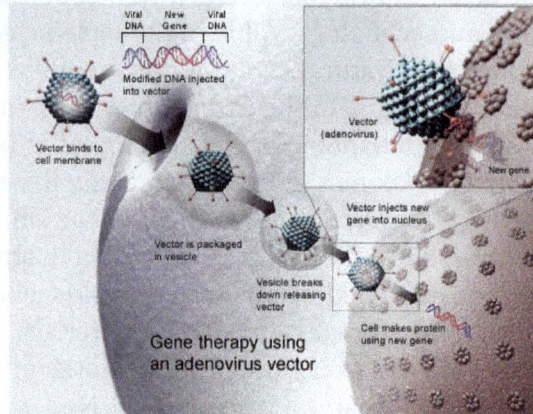

Gene therapy using an adenovirus vector. In some cases, the adenovirus will insert the new gene into a cell. If the treatment is successful, the new gene will make a functional protein to treat a disease.

Gene therapy is the therapeutic delivery of nucleic acid polymers into a patient's cells as a drug to treat disease.

The origins of gene therapy can be traced back to the first live attenuated vaccines in the 1950s. Although attenuated vaccines do not alter extant human genes, viruses are RNA polymers with their own genetic code that acts upon human cells, thus live vaccines can be considered a primitive form of gene therapy, albeit not in the sense that is generally implied today.

The first attempt at modifying human DNA was performed in 1980 by Martin Cline, but the first successful and approved nuclear gene transfer in humans was performed in May 1989. The first therapeutic use of gene transfer as well as the first direct insertion of human DNA into the nuclear genome was performed by French Anderson in a trial starting in September 1990.

Between 1989 and February 2016, over 2,300 clinical trials had been conducted, more than half of them in phase I.

It should be noted that not all medical procedures that introduce alterations to a patient's genetic makeup can be considered gene therapy. Bone marrow transplantation and organ transplants in general have been found to introduce foreign DNA into patients. Gene therapy is defined by the precision of the procedure and the intention of direct therapeutic effects.

Background

Gene therapy was conceptualized in 1972, by authors who urged caution before commencing human gene therapy studies.

The first attempt, albeit an unsuccessful one, at gene therapy (as well as the first case of medical transfer of foreign genes into humans not counting organ transplantation) was performed by Martin Cline on 10 July 1980. Cline claimed that one of the genes in his patients was active six months later, though he never published this data or had it verified and even if he is correct, it's unlikely it produced any significant beneficial effects treating beta-thalassemia.

After extensive research on animals throughout the 1980s and a 1989 bacterial gene tagging trial on humans, the first gene therapy widely accepted as a success was demonstrated in a trial that started on September 14, 1990, when Ashi DeSilva was treated for ADA-SCID.

The first somatic treatment that produced a permanent genetic change was performed in 1993.

This procedure was referred to sensationally and somewhat inaccurately in the media as a "three parent baby", though mtDNA is not the primary human genome and has little effect on an organism's individual characteristics beyond powering their cells.

Gene therapy is a way to fix a genetic problem at its source. The polymers are either translated into proteins, interfere with target gene expression, or possibly correct genetic mutations.

The most common form uses DNA that encodes a functional, therapeutic gene to replace a mutated gene. The polymer molecule is packaged within a "vector", which carries the molecule inside cells.

Early clinical failures led to dismissals of gene therapy. Clinical successes since 2006 regained researchers' attention, although as of 2014, it was still largely an experimental technique. These include treatment of retinal diseases Leber's congenital amaurosis and choroideremia, X-linked SCID, ADA-SCID, adrenoleukodystrophy, chronic lymphocytic leukemia (CLL), acute lymphocytic leukemia (ALL), multiple myeloma, haemophilia and Parkinson's disease. Between 2013 and April 2014, US companies invested over $600 million in the field.

The first commercial gene therapy, Gendicine, was approved in China in 2003 for the treatment of certain cancers. In 2011 Neovasculgen was registered in Russia as the first-in-class gene-therapy drug for treatment of peripheral artery disease, including critical limb ischemia. In 2012 Glybera, a treatment for a rare inherited disorder, became the first treatment to be approved for clinical use in either Europe or the United States after its endorsement by the European Commission.

Approaches

Following early advances in genetic engineering of bacteria, cells and small animals, scientists started considering how to apply it to medicine. Two main approaches were considered – replacing or disrupting defective genes. Scientists focused on diseases caused by single-gene defects, such as cystic fibrosis, haemophilia, muscular dystrophy, thalassemia and sickle cell anemia. Glybera treats one such disease, caused by a defect in lipoprotein lipase.

DNA must be administered, reach the damaged cells, enter the cell and express/disrupt a protein. Multiple delivery techniques have been explored. The initial approach incorporated DNA into an engineered virus to deliver the DNA into a chromosome. Naked DNA approaches have also been explored, especially in the context of vaccine development.

Generally, efforts focused on administering a gene that causes a needed protein to be expressed. More recently, increased understanding of nuclease function has led to more direct DNA editing, using techniques such as zinc finger nucleases and CRISPR. The vector incorporates genes into chromosomes. The expressed nucleases then knock out and replace genes in the chromosome. As of 2014 these approaches involve removing cells from patients, editing a chromosome and returning the transformed cells to patients.

Future of CRISPR-Cas 9

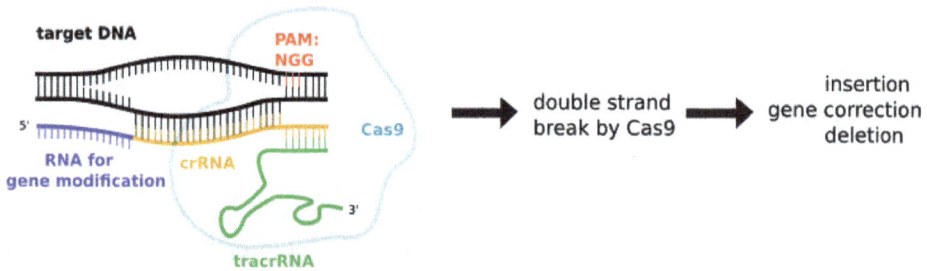

A duplex of crRNA and tracrRNA acts as guide RNA to introduce a specifically located gene modification based on the RNA 5' upstream of the crRNA. Cas9 binds the tracrRNA and needs a DNA binding sequence (5'NGG3'), which is called protospacer adjacent motif (PAM). After binding,Cas9 introduces a DNA double strand break, which is then followed by gene modification via bhomologous recombination (HDR) or non-homologous end joining (NHEJ).

Gene editing has been a potential therapy for many genetic diseases. Targeted genome editing using nucleases provides a general method for inducing deletions or insertion. An earlier method for targeting relies on protein-DNA interactions; however, the most recent one, using CRISPR – associated protein 9 (Cas9), provides better specificity, simplicity, speed and pricing. The CRISPR System was first identified in single cell archaea (Prokaryotes). It is now widely used in different cell types and organisms including human cells (HEK293T, HeLa, iPSC), mouse, fruit fly, rice, wheat etc. CRISPR –Cas9 genome editing has potential applications in human gene therapy, screening drug targets, synthetic biology, agriculture, programmable RNA targeting, and viral gene disruption.

Other technologies employ antisense, small interfering RNA and other DNA. To the extent that these technologies do not alter DNA, but instead directly interact with molecules such as RNA, they are not considered "gene therapy" per se.

Cell Types

Gene therapy may be classified into two types:

Somatic Cell

In somatic cell gene therapy (SCGT), the therapeutic genes are transferred into any cell other than a gamete, germ cell, gametocyte or undifferentiated stem cell. Any such modifications affect the individual patient only, and are not inherited by offspring. Somatic gene therapy represents mainstream basic and clinical research, in which therapeutic DNA (either integrated in the genome or as an external episome or plasmid) is used to treat disease.

Over 600 clinical trials utilizing SCGT are underway in the US. Most focus on severe genetic disorders, including immunodeficiencies, haemophilia, thalassaemia and cystic fibrosis. Such single gene disorders are good candidates for somatic cell therapy. The complete correction of a genetic disorder or the replacement of multiple genes is not yet possible. Only a few of the trials are in the advanced stages.

Germline

In germline gene therapy (GGT), germ cells (sperm or eggs) are modified by the introduction of functional genes into their genomes. Modifying a germ cell causes all the organism's cells to contain the modified gene. The change is therefore heritable and passed on to later generations. Australia, Canada, Germany, Israel, Switzerland and the Netherlands prohibit GGT for application in human beings, for technical and ethical reasons, including insufficient knowledge about possible risks to future generations and higher risks versus SCGT. The US has no federal controls specifically addressing human genetic modification (beyond FDA regulations for therapies in general).

Vectors

The delivery of DNA into cells can be accomplished by multiple methods. The two major classes are recombinant viruses (sometimes called biological nanoparticles or viral vectors) and naked DNA or DNA complexes (non-viral methods).

Viruses

In order to replicate, viruses introduce their genetic material into the host cell, tricking the host's cellular machinery into using it as blueprints for viral proteins. Scientists exploit this by substituting a virus's genetic material with therapeutic DNA. (The term 'DNA' may be an oversimplification, as some viruses contain RNA, and gene therapy could take this form as well.) A number of viruses have been used for human gene therapy, including retrovirus, adenovirus, lentivirus, herpes simplex, vaccinia and adeno-associated virus. Like the genetic material (DNA or RNA) in viruses, therapeutic DNA can be designed to simply serve as a temporary blueprint that is degraded naturally or (at least theoretically) to enter the host's genome, becoming a permanent part of the host's DNA in infected cells.

Non-viral

Non-viral methods present certain advantages over viral methods, such as large scale production and low host immunogenicity. However, non-viral methods initially produced lower levels of transfection and gene expression, and thus lower therapeutic efficacy. Later technology remedied this deficiency.

Methods for non-viral gene therapy include the injection of naked DNA, electroporation, the gene gun, sonoporation, magnetofection, the use of oligonucleotides, lipoplexes, dendrimers, and inorganic nanoparticles.

Hurdles

Some of the unsolved problems include:

- Short-lived nature – Before gene therapy can become a permanent cure for a condition, the therapeutic DNA introduced into target cells must remain functional and the cells containing the therapeutic DNA must be stable. Problems with integrating therapeutic DNA into the genome and the rapidly dividing nature of many cells prevent it from achieving long-term benefits. Patients require multiple treatments.

- Immune response – Any time a foreign object is introduced into human tissues, the immune system is stimulated to attack the invader. Stimulating the immune system in a way that reduces gene therapy effectiveness is possible. The immune system's enhanced response to viruses that it has seen before reduces the effectiveness to repeated treatments.

- Problems with viral vectors – Viral vectors carry the risks of toxicity, inflammatory responses, and gene control and targeting issues.

- Multigene disorders – Some commonly occurring disorders, such as heart disease, high blood pressure, Alzheimer's disease, arthritis, and diabetes, are affected by variations in multiple genes, which complicate gene therapy.

- Some therapies may breach the Weismann barrier (between soma and germline) protecting the testes, potentially modifying the germline, falling afoul of regulations in countries that prohibit the latter practice.

- Insertional mutagenesis – If the DNA is integrated in a sensitive spot in the genome, for example in a tumor suppressor gene, the therapy could induce a tumor. This has occurred in clinical trials for X-linked severe combined immunodeficiency (X-SCID) patients, in which hematopoietic stem cells were transduced with a corrective transgene using a retrovirus, and this led to the development of T cell leukemia in 3 of 20 patients. One possible solution is to add a functional tumor suppressor gene to the DNA to be integrated. This may be problematic since the longer the DNA is, the harder it is to integrate into cell genomes. CRISPR technology allows researchers to make much more precise genome changes at exact locations.

- Cost – Alipogene tiparvovec or Glybera, for example, at a cost of $1.6 million per patient, was reported in 2013 to be the world's most expensive drug.

Deaths

Three patients' deaths have been reported in gene therapy trials, putting the field under close scrutiny. The first was that of Jesse Gelsinger in 1999. One X-SCID patient died of leukemia in 2003. In 2007, a rheumatoid arthritis patient died from an infection; the subsequent investigation concluded that the death was not related to gene therapy.

History

1970s and Earlier

In 1972 Friedmann and Roblin authored a paper in *Science* titled "Gene therapy for human genetic disease?" Rogers (1970) was cited for proposing that *exogenous good DNA* be used to replace the defective DNA in those who suffer from genetic defects.

1980s

In 1984 a retrovirus vector system was designed that could efficiently insert foreign genes into mammalian chromosomes.

1990s

The first approved gene therapy clinical research in the US took place on 14 September 1990, at the National Institutes of Health (NIH), under the direction of William French Anderson. Four-year-old Ashanti DeSilva received treatment for a genetic defect that left her with ADA-SCID, a severe immune system deficiency. The effects were temporary, but successful.

Cancer gene therapy was introduced in 1992/93 (Trojan et al. 1993). The treatment of glioblastoma multiforme, the malignant brain tumor whose outcome is always fatal, was done using a vector expressing antisense IGF-I RNA (clinical trial approved by

NIH n° 1602, and FDA in 1994). The therapy proved to be effective due to the anti-tumor mechanism of IGF-I antisense, which is related to strong immune and apoptotic phenomena.

In 1992 Claudio Bordignon, working at the Vita-Salute San Raffaele University, performed the first gene therapy procedure using hematopoietic stem cells as vectors to deliver genes intended to correct hereditary diseases. In 2002 this work led to the publication of the first successful gene therapy treatment for adenosine deaminase-deficiency (SCID). The success of a multi-center trial for treating children with SCID (severe combined immune deficiency or "bubble boy" disease) from 2000 and 2002, was questioned when two of the ten children treated at the trial's Paris center developed a leukemia-like condition. Clinical trials were halted temporarily in 2002, but resumed after regulatory review of the protocol in the US, the United Kingdom, France, Italy and Germany.

In 1993 Andrew Gobea was born with SCID following prenatal genetic screening. Blood was removed from his mother's placenta and umbilical cord immediately after birth, to acquire stem cells. The allele that codes for adenosine deaminase (ADA) was obtained and inserted into a retrovirus. Retroviruses and stem cells were mixed, after which the viruses inserted the gene into the stem cell chromosomes. Stem cells containing the working ADA gene were injected into Andrew's blood. Injections of the ADA enzyme were also given weekly. For four years T cells (white blood cells), produced by stem cells, made ADA enzymes using the ADA gene. After four years more treatment was needed.

Jesse Gelsinger's death in 1999 impeded gene therapy research in the US. As a result, the FDA suspended several clinical trials pending the reevaluation of ethical and procedural practices.

2000s

The modified cancer gene therapy strategy of antisense IGF-I RNA (NIH n° 1602) using antisense / triple helix anti IGF-I approach was registered in 2002 by Wiley gene therapy clinical trial - n° 635 and 636. The approach has shown promising results in the treatment of six different malignant tumors: glioblastoma, cancers of liver, colon, prostate, uterus and ovary (Collaborative NATO Science Programme on Gene Therapy USA, France, Poland n° LST 980517 conducted by J. Trojan) (Trojan et al., 2012). This anti–gene antisense/triple helix therapy has proven to be efficient, due to the mechanism stopping simultaneously IGF-I expression on translation and transcription levels, strengthening anti-tumor immune and apoptotic phenomena.

2002

Sickle-cell disease can be treated in mice. The mice – which have essentially the same defect that causes human cases – used a viral vector to induce production of fetal he-moglobin (HbF), which normally ceases to be produced shortly after birth. In humans, the use of hydroxyurea to stimulate the production of HbF temporarily alleviates sickle cell symptoms. The researchers demonstrated this treatment to be a more permanent means to increase therapeutic HbF production.

A new gene therapy approach repaired errors in messenger RNA derived from defective genes. This technique has the potential to treat thalassaemia, cystic fibrosis and some cancers.

Researchers created liposomes 25 nanometers across that can carry therapeutic DNA through pores in the nuclear membrane.

2003

In 2003 a research team inserted genes into the brain for the first time. They used li-posomes coated in a polymer called polyethylene glycol, which, unlike viral vectors, are small enough to cross the blood–brain barrier.

Short pieces of double-stranded RNA (short, interfering RNAs or siRNAs) are used by cells to degrade RNA of a particular sequence. If a siRNA is designed to match the RNA copied from a faulty gene, then the abnormal protein product of that gene will not be produced.

Gendicine is a cancer gene therapy that delivers the tumor suppressor gene p53 using an engineered adenovirus. In 2003, it was approved in China for the treatment of head and neck squamous cell carcinoma.

2006

In March researchers announced the successful use of gene therapy to treat two adult patients for X-linked chronic granulomatous disease, a disease which affects myeloid cells and damages the immune system. The study is the first to show that gene therapy can treat the myeloid system.

In May a team reported a way to prevent the immune system from rejecting a newly delivered gene. Similar to organ transplantation, gene therapy has been plagued by this problem. The immune system normally recognizes the new gene as foreign and rejects the cells carrying it. The research utilized a newly uncovered network of genes regulated by molecules known as microRNAs. This natural function selectively obscured their ther-apeutic gene in immune system cells and protected it from discovery. Mice infected with the gene containing an immune-cell microRNA target sequence did not reject the gene.

In August scientists successfully treated metastatic melanoma in two patients using killer T cells genetically retargeted to attack the cancer cells.

In November researchers reported on the use of VRX496, a gene-based immunotherapy for the treatment of HIV that uses a lentiviral vector to deliver an antisense gene against the HIV envelope. In a phase I clinical trial, five subjects with chronic HIV infection who had failed to respond to at least two antiretroviral regimens were treated. A single intravenous infusion of autologous CD4 T cells genetically modified with VRX496 was well tolerated. All patients had stable or decreased viral load; four of the five patients had stable or increased CD4 T cell counts. All five patients had stable or increased immune response to HIV antigens and other pathogens. This was the first evaluation of a lentiviral vector administered in a US human clinical trial.

2007

In May researchers announced the first gene therapy trial for inherited retinal disease. The first operation was carried out on a 23-year-old British male, Robert Johnson, in early 2007.

2008

Leber's congenital amaurosis is an inherited blinding disease caused by mutations in the RPE65 gene. The results of a small clinical trial in children were published in April. Delivery of recombinant adeno-associated virus (AAV) carrying RPE65 yielded positive results. In May two more groups reported positive results in independent clinical trials using gene therapy to treat the condition. In all three clinical trials, patients recovered functional vision without apparent side-effects.

2009

In September researchers were able to give trichromatic vision to squirrel monkeys. In November 2009, researchers halted a fatal genetic disorder called adrenoleukodystrophy in two children using a lentivirus vector to deliver a functioning version of ABCD1, the gene that is mutated in the disorder.

2010s

2010

An April paper reported that gene therapy addressed achromatopsia (color blindness) in dogs by targeting cone photoreceptors. Cone function and day vision were restored for at least 33 months in two young specimens. The therapy was less efficient for older dogs.

In September it was announced that an 18-year-old male patient in France with beta-thalassemia major had been successfully treated. Beta-thalassemia major is an inherited blood disease in which beta haemoglobin is missing and patients are dependent on regular lifelong blood transfusions. The technique used a lentiviral vector to transduce the human ß-globin gene into purified blood and marrow cells obtained from the patient in June 2007. The patient's haemoglobin levels were stable at 9 to 10 g/dL. About a third of the hemoglobin contained the form introduced by the viral vector and blood transfusions were not needed. Further clinical trials were planned. Bone marrow transplants are the only cure for thalassemia, but 75% of patients do not find a matching donor.

2011

In 2007 and 2008, a man was cured of HIV by repeated Hematopoietic stem cell transplantation with double-delta-32 mutation which disables the CCR5 receptor. This cure was accepted by the medical community in 2011. It required complete ablation of existing bone marrow, which is very debilitating.

In August two of three subjects of a pilot study were confirmed to have been cured from chronic lymphocytic leukemia (CLL). The therapy used genetically modified T cells to attack cells that expressed the CD19 protein to fight the disease. In 2013, the researchers announced that 26 of 59 patients had achieved complete remission and the original patient had remained tumor-free.

Human HGF plasmid DNA therapy of cardiomyocytes is being examined as a potential treatment for coronary artery disease as well as treatment for the damage that occurs to the heart after myocardial infarction.

In 2011 Neovasculgen was registered in Russia as the first-in-class gene-therapy drug for treatment of peripheral artery disease, including critical limb ischemia; it delivers the gene encoding for VEGF. Neovasculogen is a plasmid encoding the CMV promoter and the 165 amino acid form of VEGF.

2012

The FDA approved Phase 1 clinical trials on thalassemia major patients in the US for 10 participants in July. The study was expected to continue until 2015.

In July 2012, the European Medicines Agency recommended approval of a gene therapy treatment for the first time in either Europe or the United States. The treatment used Alipogene tiparvovec (Glybera) to compensate for lipoprotein lipase deficiency, which can cause severe pancreatitis. The recommendation was endorsed by the European Commission in November 2012 and commercial rollout began in late 2014.

In December 2012, it was reported that 10 of 13 patients with multiple myeloma were in remission "or very close to it" three months after being injected with a treatment involving genetically engineered T cells to target proteins NY-ESO-1 and LAGE-1, which exist only on cancerous myeloma cells.

2013

In March researchers reported that three of five subjects who had acute lymphocytic leukemia (ALL) had been in remission for five months to two years after being treated with genetically modified T cells which attacked cells with CD19 genes on their surface, i.e. all B-cells, cancerous or not. The researchers believed that the patients' immune systems would make normal T-cells and B-cells after a couple of months. They were also given bone marrow. One patient relapsed and died and one died of a blood clot unrelated to the disease.

Following encouraging Phase 1 trials, in April, researchers announced they were starting Phase 2 clinical trials (called CUPID2 and SERCA-LVAD) on 250 patients at several hospitals to combat heart disease. The therapy was designed to increase the levels of SERCA2, a protein in heart muscles, improving muscle function. The FDA granted this a Breakthrough Therapy Designation to accelerate the trial and approval process. In 2016 it was reported that no improvement was found from the CUPID 2 trial.

In July researchers reported promising results for six children with two severe hereditary diseases had been treated with a partially deactivated lentivirus to replace a faulty gene and after 7–32 months. Three of the children had metachromatic leukodystrophy, which causes children to lose cognitive and motor skills. The other children had Wiskott-Aldrich syndrome, which leaves them to open to infection, autoimmune diseases and cancer. Follow up trials with gene therapy on another six children with Wiskott-Aldrich syndrome were also reported as promising.

In October researchers reported that two children born with adenosine deaminase severe combined immunodeficiency disease (ADA-SCID) had been treated with genetically engineered stem cells 18 months previously and that their immune systems were showing signs of full recovery. Another three children were making progress. In 2014 a further 18 children with ADA-SCID were cured by gene therapy. ADA-SCID children have no functioning immune system and are sometimes known as "bubble children."

Also in October researchers reported that they had treated six haemophilia sufferers in early 2011 using an adeno-associated virus. Over two years later all six were producing clotting factor.

Data from three trials on Topical cystic fibrosis transmembrane conductance regulator gene therapy were reported to not support its clinical use as a mist inhaled into the lungs to treat cystic fibrosis patients with lung infections.

2014

In January researchers reported that six choroideremia patients had been treated with adeno-associated virus with a copy of REP1. Over a six-month to two-year period all had improved their sight. By 2016, 32 patients had been treated with positive results and researchers were hopeful the treatment would be long-lasting. Choroideremia is an inherited genetic eye disease with no approved treatment, leading to loss of sight.

In March researchers reported that 12 HIV patients had been treated since 2009 in a trial with a genetically engineered virus with a rare mutation (CCR5 deficiency) known to protect against HIV with promising results.

Clinical trials of gene therapy for sickle cell disease were started in 2014 although one review failed to find any such trials.

2015

In February LentiGlobin BB305, a gene therapy treatment undergoing clinical trials for treatment of beta thalassemia gained FDA "breakthrough" status after several patients were able to forgo the frequent blood transfusions usually required to treat the disease.

In March researchers delivered a recombinant gene encoding a broadly neutralizing antibody into monkeys infected with simian HIV; the monkeys' cells produced the antibody, which cleared them of HIV. The technique is named immunoprophylaxis by gene transfer (IGT). Animal tests for antibodies to ebola, malaria, influenza and hepatitis are underway.

In March scientists, including an inventor of CRISPR, urged a worldwide moratorium on germline gene therapy, writing "scientists should avoid even attempting, in lax jurisdictions, germline genome modification for clinical application in humans" until the full implications "are discussed among scientific and governmental organizations".

Also in 2015 Glybera was approved for the German market.

In October, researchers announced that they had treated a baby girl, Layla Richards, with an experimental treatment using donor T-cells genetically engineered to attack cancer cells. Two months after the treatment she was still free of her cancer (a highly aggressive form of acute lymphoblastic leukaemia [ALL]). Children with highly aggressive ALL normally have a very poor prognosis and Layla's disease had been regarded as terminal before the treatment.

In December, scientists of major world academies called for a moratorium on inheritable human genome edits, including those related to CRISPR-Cas9 technologies but that basic research including embryo gene editing should continue.

2016

In April the Committee for Medicinal Products for Human Use of the European Medicines Agency endorsed a gene therapy treatment called Strimvelis and recommended it be approved. This treats children born with ADA-SCID and who have no functioning immune system - sometimes called the "bubble baby" disease. This would be the second gene therapy treatment to be approved in Europe.

Speculative Uses

Speculated uses for gene therapy include:

Fertility

Gene Therapy techniques have the potential to provide alternative treatments for those with infertility. Recently, successful experimentation on mice has proven that fertility can be restored by using the gene therapy method, CRISPR. Spermatogenical stem cells from another organism were transplanted into the testes of an infertile male mouse. The stem cells re-established spermatogenesis and fertility.

Gene Doping

Athletes might adopt gene therapy technologies to improve their performance. Gene doping is not known to occur, but multiple gene therapies may have such effects. Kayser et al. argue that gene doping could level the playing field if all athletes receive equal access. Critics claim that any therapeutic intervention for non-therapeutic/enhancement purposes compromises the ethical foundations of medicine and sports.

Human Genetic Engineering

Genetic engineering could be used to change physical appearance, metabolism, and even improve physical capabilities and mental faculties such as memory and intelligence. Ethical claims about germline engineering include beliefs that every fetus has a right to remain genetically unmodified, that parents hold the right to genetically modify their offspring, and that every child has the right to be born free of preventable diseases. For adults, genetic engineering could be seen as another enhancement technique to add to diet, exercise, education, cosmetics and plastic surgery. Another theorist claims that moral concerns limit but do not prohibit germline engineering.

Possible regulatory schemes include a complete ban, provision to everyone, or professional self-regulation. The American Medical Association's Council on Ethical and Judicial Affairs stated that "genetic interventions to enhance traits should be considered permissible only in severely restricted situations: (1) clear and meaningful benefits to the fetus or child; (2) no trade-off with other characteristics or traits; and (3) equal access to the genetic technology, irrespective of income or other socioeconomic characteristics."

As early in the history of biotechnology as 1990, there have been scientists opposed to attempts to modify the human germline using these new tools, and such concerns have continued as technology progressed. With the advent of new techniques like CRISPR, in March 2015 a group of scientists urged a worldwide moratorium on clinical use of gene editing technologies to edit the human genome in a way that can be inherited. In April 2015, researchers sparked controversy when they reported results of basic research to edit the DNA of non-viable human embryos using CRISPR.

Regulations

Regulations covering genetic modification are part of general guidelines about human-involved biomedical research.

The Helsinki Declaration (Ethical Principles for Medical Research Involving Human Subjects) was amended by the World Medical Association's General Assembly in 2008. This document provides principles physicians and researchers must consider when involving humans as research subjects. The Statement on Gene Therapy Research initiated by the Human Genome Organization (HUGO) in 2001 provides a legal baseline for all countries. HUGO's document emphasizes human freedom and adherence to human rights, and offers recommendations for somatic gene therapy, including the importance of recognizing public concerns about such research.

United States

No federal legislation lays out protocols or restrictions about human genetic engineering. This subject is governed by overlapping regulations from local and federal agencies, including the Department of Health and Human Services, the FDA and NIH's Recombinant DNA Advisory Committee. Researchers seeking federal funds for an investigational new drug application, (commonly the case for somatic human genetic engineering), must obey international and federal guidelines for the protection of human subjects.

NIH serves as the main gene therapy regulator for federally funded research. Privately funded research is advised to follow these regulations. NIH provides funding for research that develops or enhances genetic engineering techniques and to evaluate the ethics and quality in current research. The NIH maintains a mandatory registry of human genetic engineering research protocols that includes all federally funded projects.

An NIH advisory committee published a set of guidelines on gene manipulation. The guidelines discuss lab safety as well as human test subjects and various experimental types that involve genetic changes. Several sections specifically pertain to human genetic engineering, including Section III-C-1. This section describes required review processes and other aspects when seeking approval to begin clinical research involving genetic transfer into a human patient. The protocol for a gene therapy clinical trial must be approved by the NIH's Recombinant DNA Advisory Committee prior to any clinical trial beginning; this is different from any other kind of clinical trial.

As with other kinds of drugs, the FDA regulates the quality and safety of gene therapy products and supervises how these products are used clinically. Therapeutic alteration of the human genome falls under the same regulatory requirements as any other medical treatment. Research involving human subjects, such as clinical trials, must be reviewed and approved by the FDA and an Institutional Review Board.

Popular Culture

Gene therapy is the basis for the plotline of the film *I Am Legend* and the TV show *Will Gene Therapy Change the Human Race?*.

Virology

Virology is the study of viruses – submicroscopic, parasitic particles of genetic material contained in a protein coat – and virus-like agents. It focuses on the following aspects of viruses: their structure, classification and evolution, their ways to infect and exploit host cells for reproduction, their interaction with host organism physiology and immunity, the diseases they cause, the techniques to isolate and culture them, and their use in research and therapy. Virology is considered to be a subfield of microbiology or of medicine.

Virus Structure and Classification

A major branch of virology is virus classification. Viruses can be classified according to the host cell they infect: animal viruses, plant viruses, fungal viruses, and bacteriophages (viruses infecting bacterium, which include the most complex viruses). Another classification uses the geometrical shape of their capsid (often a helix or an icosahedron) or the virus's structure (e.g. presence or absence of a lipid envelope). Viruses range in size from about 30 nm to about 450 nm, which means that most of them cannot be seen with light microscopes. The shape and structure of viruses has been studied by electron microscopy, NMR spectroscopy, and X-ray crystallography.

The most useful and most widely used classification system distinguishes viruses according to the type of nucleic acid they use as genetic material and the viral replication method they employ to coax host cells into producing more viruses:

- DNA viruses (divided into double-stranded DNA viruses and single-stranded DNA viruses),

- RNA viruses (divided into positive-sense single-stranded RNA viruses, negative-sense single-stranded RNA viruses and the much less common double-stranded RNA viruses),

- reverse transcribing viruses (double-stranded reverse-transcribing DNA viruses and single-stranded reverse-transcribing RNA viruses including retroviruses).

The latest report by the International Committee on Taxonomy of Viruses (2005) lists 5450 viruses, organized in over 2,000 species, 287 genera, 73 families and 3 orders.

Virologists also study *subviral particles*, infectious entities notably smaller and simpler than viruses:

- viroids (naked circular RNA molecules infecting plants),

- satellites (nucleic acid molecules with or without a capsid that require a helper virus for infection and reproduction), and

- prions (proteins that can exist in a pathological conformation that induces other prion molecules to assume that same conformation).

Taxa in virology are not necessarily monophyletic, as the evolutionary relationships of the various virus groups remain unclear. Three hypotheses regarding their origin exist:

1. Viruses arose from non-living matter, separately from yet in parallel to cells, perhaps in the form of self-replicating RNA ribozymes similar to viroids.

2. Viruses arose by genome reduction from earlier, more competent cellular life forms that became parasites to host cells and subsequently lost most of their functionality; examples of such tiny parasitic prokaryotes are Mycoplasma and Nanoarchaea.

3. Viruses arose from mobile genetic elements of cells (such as transposons, retrotransposons or plasmids) that became encapsulated in protein capsids, acquired the ability to "break free" from the host cell and infect other cells.

Of particular interest here is mimivirus, a giant virus that infects amoebae and encodes much of the molecular machinery traditionally associated with bacteria. Is it a simplified version of a parasitic prokaryote, or did it originate as a simpler virus that acquired genes from its host?

The evolution of viruses, which often occurs in concert with the evolution of their hosts, is studied in the field of viral evolution.

While viruses reproduce and evolve, they do not engage in metabolism, do not move, and depend on a host cell for reproduction. The often-debated question of whether they are alive or not is a matter of definition that does not affect the biological reality of viruses.

Viral Diseases and Host Defenses

One main motivation for the study of viruses is the fact that they cause many important infectious diseases, among them the common cold, influenza, rabies, measles, many forms of diarrhea, hepatitis, Dengue fever, yellow fever, polio, smallpox and AIDS. Herpes simplex causes cold sores and genital herpes and is under investigation as a possible factor in Alzheimer's.

Some viruses, known as oncoviruses, contribute to the development of certain forms of cancer. The best studied example is the association between Human papillomavirus and cervical cancer: almost all cases of cervical cancer are caused by certain strains of this sexually transmitted virus. Another example is the association of infection with hepatitis B and hepatitis C viruses and liver cancer.

Some subviral particles also cause disease: the transmissible spongiform encephalopathies, which include Kuru, Creutzfeldt–Jakob disease and bovine spongiform encephalopathy ("mad cow disease"), are caused by prions, hepatitis D is due to a satellite virus.

The study of the manner in which viruses cause disease is viral pathogenesis. The degree to which a virus causes disease is its virulence.

When the immune system of a vertebrate encounters a virus, it may produce specific antibodies which bind to the virus and neutralize its infectivity or mark it for destruction. Antibody presence in blood serum is often used to determine whether a person has been exposed to a given virus in the past, with tests such as ELISA. Vaccinations protect against viral diseases, in part, by eliciting the production of antibodies. Monoclonal antibodies, specific to the virus, are also used for detection, as in fluorescence microscopy.

A second defense of vertebrates against viruses, cell-mediated immunity, involves immune cells known as T cells: the body's cells constantly display short fragments of their proteins on the cell's surface, and if a T cell recognizes a suspicious viral fragment there, the host cell is destroyed and the virus-specific T-cells proliferate. This mechanism is jump-started by certain vaccinations.

RNA interference, an important cellular mechanism found in plants, animals and many other eukaryotes, most likely evolved as a defense against viruses. An elaborate machinery of interacting enzymes detects double-stranded RNA molecules (which occur as part of the life cycle of many viruses) and then proceeds to destroy all single-stranded versions of those detected RNA molecules.

Every lethal viral disease presents a paradox: killing its host is obviously of no benefit to the virus, so how and why did it evolve to do so? Today it is believed that most viruses are relatively benign in their natural hosts; some viral infection might even be beneficial to the host. The lethal viral diseases are believed to have resulted from an "accidental" jump of the virus from a species in which it is benign to a new one that is not accustomed to it. For example, viruses that cause serious influenza in humans probably have pigs or birds as their natural host, and HIV is thought to derive from the benign non-human primate virus SIV.

While it has been possible to prevent (certain) viral diseases by vaccination for a long time, the development of antiviral drugs to *treat* viral diseases is a comparatively recent development. The first such drug was interferon, a substance that is naturally produced when an infection is detected and stimulates other parts of the immune system.

Molecular Biology Research and Viral Therapy

Bacteriophages, the viruses which infect bacteria, can be relatively easily grown as viral plaques on bacterial cultures. Bacteriophages occasionally move genetic material from one bacterial cell to another in a process known as transduction, and this horizontal gene transfer is one reason why they served as a major research tool in the early development of molecular biology. The genetic code, the function of ribozymes, the first recombinant DNA and early genetic libraries were all arrived at using bacteriophages. Certain genetic elements derived from viruses, such as highly effective promoters, are commonly used in molecular biology research today.

Growing animal viruses outside of the living host animal is more difficult. Classically, fertilized chicken eggs have often been used, but cell cultures are increasingly employed for this purpose today.

Since some viruses that infect eukaryotes need to transport their genetic material into the host cell's nucleus, they are attractive tools for introducing new genes into the host (known as transformation or transfection). Modified retroviruses are often used for this purpose, as they integrate their genes into the host's chromosomes.

This approach of using viruses as gene vectors is being pursued in the gene therapy of genetic diseases. An obvious problem to be overcome in viral gene therapy is the rejection of the transforming virus by the immune system.

Phage therapy, the use of bacteriophages to combat bacterial diseases, was a popular research topic before the advent of antibiotics and has recently seen renewed interest.

Oncolytic viruses are viruses that preferably infect cancer cells. While early efforts to employ these viruses in the therapy of cancer failed, there have been reports in 2005 and 2006 of encouraging preliminary results.

Other Uses of Viruses

A new application of genetically engineered viruses in nanotechnology was recently described; see the uses of viruses in material science and nanotechnology. For a use in mapping neurons see the applications of pseudorabies in neuroscience.

History of Virology

A very early form of vaccination known as variolation was developed several thousand years ago in China. It involved the application of materials from smallpox sufferers in order to immunize others. In 1717 Lady Mary Wortley Montagu observed the practice in Istanbul and attempted to popularize it in Britain, but encountered considerable resistance. In 1796 Edward Jenner developed a much safer method, using cowpox to successfully immunize a young boy against smallpox, and this practice was widely adopted. Vaccinations against other viral diseases followed, including the successful rabies vaccination by Louis Pasteur in 1886. The nature of viruses however was not clear to these researchers.

Dmitri Ivanovsky

Martinus Beijerinck

In 1892, the Russian biologist Dmitry Ivanovsky used a Chamberland filter to try to isolate the bacteria that caused tobacco mosaic disease. His experiments showed that crushed leaf extracts from infected tobacco plants remained infectious after filtration. Ivanovsky reported a minuscule infectious agent or toxin, capable of passing the filter, may be being produced by a bacterium.

In 1898 Martinus Beijerinck repeated Ivanovski's work but went further and passed the "filterable agent" from plant to plant, found the action undiminished, and concluded it infectious—replicating in the host—and thus not a mere toxin. He called it *contagium vivum fluidum*. The question of whether the agent was a "living fluid" or a particle was however still open.

In 1903 it was suggested for the first time that transduction by viruses might cause cancer. In 1908 Bang and Ellerman showed that a filterable virus could transmit chicken leukemia, data largely ignored till the 1930s when leukemia became regarded as cancerous. In 1911 Peyton Rous reported the transmission of chicken sarcoma, a solid tumor, with a virus, and thus Rous became "father of tumor virology". The virus was later called Rous sarcoma virus 1 and understood to be a retrovirus. Several other cancer-causing retroviruses have since been described.

The existence of viruses that infect bacteria (bacteriophages) was first recognized by Frederick Twort in 1911, and, independently, by Félix d'Herelle in 1917. As bacteria could be grown easily in culture, this led to an explosion of virology research.

The cause of the devastating Spanish flu pandemic of 1918 was initially unclear. In late 1918, French scientists showed that a "filter-passing virus" could transmit the disease to people and animals, fulfilling Koch's postulates.

In 1926 it was shown that scarlet fever is caused by a bacterium that is infected by a certain bacteriophage.

While plant viruses and bacteriophages can be grown comparatively easily, animal viruses normally require a living host animal, which complicates their study immensely. In 1931 it was shown that influenza virus could be grown in fertilized chicken eggs, a method that is still used today to produce vaccines. In 1937, Max Theiler managed to grow the yellow fever virus in chicken eggs and produced a vaccine from an attenuated virus strain; this vaccine saved millions of lives and is still being used today.

Max Delbrück, an important investigator in the area of bacteriophages, described the basic "life cycle" of a virus in 1937: rather than "growing", a virus particle is assembled from its constituent pieces in one step; eventually it leaves the host cell to infect other cells. The Hershey–Chase experiment in 1952 showed that only DNA and not protein enters a bacterial cell upon infection with bacteriophage T2. Transduction of bacteria by bacteriophages was first described in the same year.

In 1949 John F. Enders, Thomas Weller and Frederick Robbins reported growth of poliovirus in cultured human embryonal cells, the first significant example of an animal virus grown outside of animals or chicken eggs. This work aided Jonas Salk in deriving a polio vaccine from deactivated polio viruses; this vaccine was shown to be effective in 1955.

The first virus that could be crystalized and whose structure could therefore be elucidated in detail was tobacco mosaic virus (TMV), the virus that had been studied earlier by Ivanovski and Beijerink. In 1935, Wendell Stanley achieved its crystallization for electron microscopy and showed that it remains active even after crystallization. Clear X-ray diffraction pictures of the crystallized virus were obtained by Bernal and Fankuchen in 1941. Based on such pictures, Rosalind Franklin proposed the full structure of the tobacco mosaic virus in 1955. Also in 1955, Heinz Fraenkel-Conrat and Robley Williams showed that purified TMV RNA and its capsid (coat) protein can self-assemble into functional virions, suggesting that this assembly mechanism is also used within the host cell, as Delbrück had proposed earlier.

In 1963, the Hepatitis B virus was discovered by Baruch Blumberg who went on to develop a hepatitis B vaccine.

In 1965, Howard Temin described the first retrovirus: a virus whose RNA genome was reverse transcribed into complementary DNA (cDNA), then integrated into the host's genome and expressed from that template. The viral enzyme reverse transcriptase, which along with integrase is a distinguishing trait of retroviruses, was first described in 1970, independently by Howard Temin and David Baltimore. The first retrovirus infecting humans was identified by Robert Gallo in 1974. Later it was found that reverse transcriptase is not specific to retroviruses; retrotransposons which code for reverse transcriptase are abundant in the genomes of all eukaryotes. About 10-40% of the human genome derives from such retrotransposons.

In 1975 the functioning of oncoviruses was clarified considerably. Until that time, it was thought that these viruses carried certain genes called oncogenes which, when inserted into the host's genome, would cause cancer. Michael Bishop and Harold Varmus showed that the oncogene of Rous sarcoma virus is in fact not specific to the virus but is contained in the genome of healthy animals of many species. The oncovirus can switch this pre-existing benign proto-oncogene on, turning it into a true oncogene that causes cancer.

1976 saw the first recorded outbreak of Ebola virus disease, a highly lethal virally transmitted disease.

In 1977, Frederick Sanger achieved the first complete sequencing of the genome of any organism, the bacteriophage Phi X 174. In the same year, Richard Roberts and Phillip Sharp independently showed that the genes of adenovirus contain introns and therefore require gene splicing. It was later realized that almost all genes of eukaryotes have introns as well.

A worldwide vaccination campaign led by the UN World Health Organization resulted in the eradication of smallpox in 1979.

In 1982, Stanley Prusiner discovered prions and showed that they cause scrapie.

The first cases of AIDS were reported in 1981, and HIV, the retrovirus causing it, was identified in 1983 by Luc Montagnier, Françoise Barré-Sinoussi and Robert Gallo. Tests detecting HIV infection by detecting the presence of HIV antibody were developed. Subsequent tremendous research efforts turned HIV into the best studied virus. Human Herpes Virus 8, the cause of Kaposi's sarcoma which is often seen in AIDS patients, was identified in 1994. Several antiretroviral drugs were developed in the late 1990s, decreasing AIDS mortality dramatically in developed countries.

The Hepatitis C virus was identified using novel molecular cloning techniques in 1987, leading to screening tests that dramatically reduced the incidence of post-transfusion hepatitis.

The first attempts at gene therapy involving viral vectors began in the early 1980s, when retroviruses were developed that could insert a foreign gene into the host's genome. They contained the foreign gene but did not contain the viral genome and therefore could not reproduce. Tests in mice were followed by tests in humans, beginning in 1989. The first human studies attempted to correct the genetic disease severe combined immunodeficiency (SCID), but clinical success was limited. In the period from 1990 to 1995, gene therapy was tried on several other diseases and with different viral vectors, but it became clear that the initially high expectations were overstated. In 1999 a further setback occurred when 18-year-old Jesse Gelsinger died in a gene therapy trial. He suffered a severe immune response after having received an adenovirus vector. Success in the gene therapy of two cases of X-linked SCID was reported in 2000.

In 2002 it was reported that poliovirus had been synthetically assembled in the laboratory, representing the first synthetic organism. Assembling the 7741-base genome from scratch, starting with the virus's published RNA sequence, took about two years. In 2003 a faster method was shown to assemble the 5386-base genome of the bacteriophage Phi X 174 in 2 weeks.

The giant mimivirus, in some sense an intermediate between tiny prokaryotes and ordinary viruses, was described in 2003 and sequenced in 2004.

The strain of Influenza A virus subtype H1N1 that killed up to 50 million people during the Spanish flu pandemic in 1918 was reconstructed in 2005. Sequence information was pieced together from preserved tissue samples of flu victims; viable virus was then synthesized from this sequence. The 2009 flu pandemic involved another strain of Influenza A H1N1, commonly known as "swine flu".

By 1985, Harald zur Hausen had shown that two strains of Human papillomavirus (HPV) cause most cases of cervical cancer. Two vaccines protecting against these strains were released in 2006.

In 2006 and 2007 it was reported that introducing a small number of specific transcription factor genes into normal skin cells of mice or humans can turn these cells into pluripotent stem cells, known as induced pluripotent stem cells. The technique uses modified retroviruses to transform the cells; this is a potential problem for human therapy since these viruses integrate their genes at a random location in the host's genome, which can interrupt other genes and potentially causes cancer.

In 2008, Sputnik virophage was described, the first known *virophage*: it uses the machinery of a helper virus to reproduce and inhibits reproduction of that helper virus. Sputnik reproduces in amoeba infected by mamavirus, a relative of the mimivirus mentioned above and the largest known virus to date.

An endogenous retrovirus (ERV) is a retrovirus whose genome has been permanently incorporated into the germ-line genome of some organism and that is therefore copied with each reproduction of that organism. It is estimated that about 9 percent of the human genome have their origin in ERVs. In 2015 it was shown that proteins from an ERV are actively expressed in 3-day-old human embryos and appear to play a role in embryonal development and protect embryos from infection by other viruses.

References

- Ehud Gazit, Plenty of room for biology at the bottom: An introduction to bionanotechnology. Imperial College Press, 2007, ISBN 978-1-86094-677-6

- Strachnan, T.; Read, A. P. (2004). Human Molecular Genetics (3rd ed.). Garland Publishing. p. 616. ISBN 0815341849.

- Sussman, Max; Topley, W. W. C.; Wilson, Graham K.; Collier, L. H.; Balows, Albert (1998). Topley & Wilson's microbiology and microbial infections. London: Arnold. p. 3. ISBN 0-340-66316-2.

- Ghosh, Pallab (28 April 2016). "Gene therapy reverses sight loss and is long-lasting". BBC News, Science & Environment. Retrieved 29 April 2016.

- "Summary of opinion1 (initial authorisation) Strimvelis" (PDF). European Medicines Agency. 1 April 2016. pp. 1–2. Retrieved 13 April 2016.

- Hirscheler, Ben (1 April 2016). "Europe gives green light to first gene therapy for children". Reuters. Retrieved 13 April 2016.

- National Institutes of Health. NIH Guidelines for Research Involving Recombinant or Synthetic Nucleic Acid Molecules. Revised April, 2016.

- BURGER, LUDWIG; HIRSCHLER, BEN (November 26, 2014). "First gene therapy drug sets million-euro price record". Reuters. Retrieved March 2015.

- Gallagher, James (21 April 2015) Gene therapy: 'Tame HIV' used to cure disease BBC News, Health, Retrieved 21 April 2015.

- "Ten things you might have missed Monday from the world of business". Boston Globe. 3 February 2015. Retrieved 13 February 2015.

- Wade, Nicholas (19 March 2015). "Scientists Seek Ban on Method of Editing the Human Genome". New York Times. Retrieved 20 March 2015.

- Sample, Ian (5 November 2015). "Baby girl is first in the world to be treated with 'designer immune cells'". The Guardian. Retrieved 6 November 2015.

- Wade, Nicholas (3 December 2015). "Scientists Place Moratorium on Edits to Human Genome

That Could Be Inherited". New York Times. Retrieved 3 December 2015.

- Walsh, Fergus (3 December 2015). "Gene editing: Is era of designer humans getting closer?". BBC News Health. Retrieved 31 December 2015.

- "Gene Therapy Turns Several Leukemia Patients Cancer Free. Will It Work for Other Cancers, Too?". Singularity Hub. Retrieved 7 January 2014.

- "Chiesi and uniQure delay Glybera launch to add data". Biotechnology. The Pharma Letter. 4 August 2014. Retrieved 28 August 2014.

- Bosely, Sarah (30 April 2013) Pioneering gene therapy trials offer hope for heart patients The Guardian. Retrieved 28 April 2014.

- Geddes, Linda (30 October 2013) 'Bubble kid' success puts gene therapy back on track' The New Scientist. Retrieved 2 November 2013.

- Coghlan, Andy (26 March 2013) Gene therapy cures leukaemia in eight days. The New Scientist. Retrieved 15 April 2013.

3

Technological Breakthrough of Biomedicine

The chapter serves as a source to understand the major technological breakthroughs of biomedicine. The major ideas dealt are health care, genome and genome project, human metabolome database etc. An informative account is given about health care and the prevention of disease, illness and injury. The themes discussed in the chapter are of great importance to broaden the existing knowledge on biomedicine.

Health Care

Health care or healthcare is the maintenance or improvement of health via the diagnosis, treatment, and prevention of disease, illness, injury, and other physical and mental impairments in human beings. Health care is delivered by health professionals (providers or practitioners) in allied health professions, chiropractic, physicians, physician associates, dentistry, midwifery, nursing, medicine, optometry, pharmacy, psychology, and other health professions. It includes the work done in providing primary care, secondary care, and tertiary care, as well as in public health.

Weill-Cornell New York-Presbyterian Hospital, white complex at centre, one of the world's busiest

Access to health care varies across countries, groups, and individuals, largely influenced by social and economic conditions as well as the health policies in place. Countries and jurisdictions have different policies and plans in relation to the personal and population-based health care goals within their societies. Health care systems are organizations established to meet the health needs of target populations. Their exact configuration varies between national and subnational entities. In some countries and jurisdictions, health care planning is distributed among market participants, whereas in others, planning occurs more centrally among governments or other coordinating bodies. In all cases, according to the World Health Organization (WHO), a well-functioning health care system requires a robust financing mechanism; a well-trained and adequately-paid workforce; reliable information on which to base decisions and policies; and well maintained health facilities and logistics to deliver quality medicines and technologies.

Health care can contribute to a significant part of a country's economy. In 2011, the health care industry consumed an average of 9.3 percent of the GDP or US$ 3,322 (PPP-adjusted) per capita across the 34 members of OECD countries. The USA (17.7%, or US$ PPP 8,508), the Netherlands (11.9%, 5,099), France (11.6%, 4,118), Germany (11.3%, 4,495), Canada (11.2%, 5669), and Switzerland (11%, 5,634) were the top spenders, however life expectancy in total population at birth was highest in Switzerland (82.8 years), Japan and Italy (82.7), Spain and Iceland (82.4), France (82.2) and Australia (82.0), while OECD's average exceeds 80 years for the first time ever in 2011: 80.1 years, a gain of 10 years since 1970. The USA (78.7 years) ranges only on place 26 among the 34 OECD member countries, but has the highest costs by far. All OECD countries have achieved universal (or almost universal) health coverage, except Mexico and the USA.

Health care is conventionally regarded as an important determinant in promoting the general physical and mental health and well-being of people around the world. An example of this was the worldwide eradication of smallpox in 1980, declared by the WHO as the first disease in human history to be completely eliminated by deliberate health care interventions.

Health Care Delivery

Primary care may be provided in community health centres.

The delivery of modern health care depends on groups of trained professionals and paraprofessionals coming together as interdisciplinary teams. This includes professionals in medicine, psychology, physiotherapy, nursing, dentistry, midwifery and allied health, plus many others such as public health practitioners, community health workers and assistive personnel, who systematically provide personal and population-based preventive, curative and rehabilitative care services.

While the definitions of the various types of health care vary depending on the different cultural, political, organizational and disciplinary perspectives, there appears to be some consensus that primary care constitutes the first element of a continuing health care process, that may also include the provision of secondary and tertiary levels of care. Healthcare can be defined as either public or private.

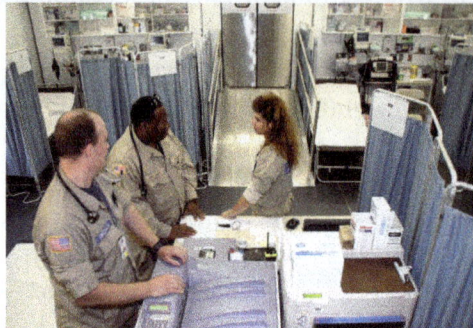

The emergency room is often a frontline venue for the delivery of primary medical care.

Primary Care

Medical train "Therapist Matvei Mudrov" in Khabarovsk, Russia

Primary care refers to the work of health professionals who act as a first point of consultation for all patients within the health care system. Such a professional would usually be a primary care physician, such as a general practitioner or family physician, a licensed independent practitioner such as a physiotherapist, or a non-physician primary care provider (mid-level provider) such as a physician assistant or nurse practitioner. Depending on the locality, health system organization, and sometimes at the patient's discretion, they may see another health care professional first, such as a pharmacist, a nurse (such as in the United Kingdom), a clinical officer (such as in parts of Africa), or an Ayurvedic or other traditional medicine professional (such as in parts of Asia). Depending on the nature of the health condition, patients may then be referred for secondary or tertiary care.

Primary care is often used as the term for the health care services which play a role in the local community. It can be provided in different settings, such as Urgent care centres which provide services to patients same day with appointment or walk-in bases.

Primary care involves the widest scope of health care, including all ages of patients, patients of all socioeconomic and geographic origins, patients seeking to maintain optimal health, and patients with all manner of acute and chronic physical, mental and social health issues, including multiple chronic diseases. Consequently, a primary care practitioner must possess a wide breadth of knowledge in many areas. Continuity is a key characteristic of primary care, as patients usually prefer to consult the same practitioner for routine check-ups and preventive care, health education, and every time they require an initial consultation about a new health problem. The International Classification of Primary Care (ICPC) is a standardized tool for understanding and analyzing information on interventions in primary care by the reason for the patient visit.

Common chronic illnesses usually treated in primary care may include, for example: hypertension, diabetes, asthma, COPD, depression and anxiety, back pain, arthritis or thyroid dysfunction. Primary care also includes many basic maternal and child health care services, such as family planning services and vaccinations. In the United States, the 2013 National Health Interview Survey found that skin disorders (42.7%), osteoarthritis and joint disorders (33.6%), back problems (23.9%), disorders of lipid metabolism (22.4%), and upper respiratory tract disease (22.1%, excluding asthma) were the most common reasons for accessing a physician.

In the United States, primary care physicians have begun to deliver primary care outside of the managed care (insurance-billing) system through direct primary care which is a subset of the more familiar concierge medicine. Physicians in this model bill patients directly for services, either on a pre-paid monthly, quarterly, or annual basis, or bill for each service in the office. Examples of direct primary care practices include Foundation Health in Colorado and Qliance in Washington.

In context of global population aging, with increasing numbers of older adults at greater risk of chronic non-communicable diseases, rapidly increasing demand for primary care services is expected in both developed and developing countries. The World Health Organization attributes the provision of essential primary care as an integral component of an inclusive primary health care strategy.

Secondary Care

Secondary care is the health care services provided by medical specialists, dental specialists and other health professionals who generally do not have first contact with patients: for example, cardiologists, urologists, endodontists, and oral and maxillofacial surgeons.

It includes acute care: necessary treatment for a short period of time for a brief but serious illness, injury or other health condition, such as in a hospital emergency department. It also includes skilled attendance during childbirth, intensive care, and medical imaging services.

The term "secondary care" is sometimes used synonymously with "hospital care". However, many secondary care providers do not necessarily work in hospitals, such as psychiatrists, clinical psychologists, occupational therapists, most dental specialties or physiotherapists (physiotherapists are also primary care providers, and a referral is not required to see a physiotherapist), and some primary care services are delivered within hospitals. Depending on the organization and policies of the national health system, patients may be required to see a primary care provider for a referral before they can access secondary care.

For example, in the United States, which operates under a mixed market health care system, some physicians might voluntarily limit their practice to secondary care by requiring patients to see a primary care provider first, or this restriction may be imposed under the terms of the payment agreements in private or group health insurance plans. In other cases medical specialists may see patients without a referral, and patients may decide whether self-referral is preferred.

In the United Kingdom and Canada, patient self-referral to a medical specialist for secondary care is rare as prior referral from another physician (either a primary care physician or another specialist) is considered necessary, regardless of whether the funding is from private insurance schemes or national health insurance.

Allied health professionals, such as physical therapists, respiratory therapists, occupational therapists, speech therapists, and dietitians, also generally work in secondary care, accessed through either patient self-referral or through physician referral.

Tertiary Care

The National Hospital for Neurology and Neurosurgery in London, United Kingdom is a specialist neurological hospital.

Tertiary care is specialized consultative health care, usually for inpatients and on referral from a primary or secondary health professional, in a facility that has personnel and facilities for advanced medical investigation and treatment, such as a tertiary referral hospital.

Examples of tertiary care services are cancer management, neurosurgery, cardiac surgery, plastic surgery, treatment for severe burns, advanced neonatology services, palliative, and other complex medical and surgical interventions.

Quaternary Care

The term quaternary care is sometimes used as an extension of tertiary care in reference to advanced levels of medicine which are highly specialized and not widely accessed. Experimental medicine and some types of uncommon diagnostic or surgical procedures are considered quaternary care. These services are usually only offered in a limited number of regional or national health care centres. This term is more prevalent in the United Kingdom, but just as applicable in the United States. A quaternary care hospital may have virtually any procedure available, whereas a tertiary care facility may not offer a sub-specialist with that training.

Home and Community Care

Many types of health care interventions are delivered outside of health facilities. They include many interventions of public health interest, such as food safety surveillance, distribution of condoms and needle-exchange programmes for the prevention of transmissible diseases.

They also include the services of professionals in residential and community settings in support of self care, home care, long-term care, assisted living, treatment for substance use disorders and other types of health and social care services.

Community rehabilitation services can assist with mobility and independence after loss of limbs or loss of function. This can include prosthesis, orthotics or wheelchairs.

Many countries, especially in the west are dealing with aging populations, and one of the priorities of the health care system is to help seniors live full, independent lives in the comfort of their own homes. There is an entire section of health care geared to providing seniors with help in day-to-day activities at home, transporting them to doctor's appointments, and many other activities that are so essential for their health and well-being. Although they provide home care for older adults in cooperation, family members and care workers may harbor diverging attitudes and values towards their joint efforts. This state of affairs presents a challenge for the design of ICT for home care.

With obesity in children rapidly becoming a major concern, health services often set up programs in schools aimed at educating children in good eating habits; making physical education compulsory in school; and teaching young adolescents to have positive self-image.

Ratings

Health care ratings are ratings or evaluations of health care used to evaluate process of care, healthcare structures and/or outcomes of a healthcare services. This information is translated into report cards that are generated by quality organizations, nonprofit, consumer groups and media. This evaluation of quality can be based on:

- Measures of Hospital quality
- Measures of Health Plan Quality
- Measures of Physician Quality
- Measures of Quality for Other Health Professionals
- Measures of Patient Experience

Related Sectors

Health care extends beyond the delivery of services to patients, encompassing many related sectors, and set within a bigger picture of financing and governance structures.

Health System

A health system, also sometimes referred to as health care system or healthcare system is the organization of people, institutions, and resources to deliver health care services to meet the health needs of target populations.

Health Care Industry

A group of Chilean 'Damas de Rojo' volunteering at their local hospital.

The health care industry incorporates several sectors that are dedicated to providing health care services and products. As a basic framework for defining the sector, the United Nations' International Standard Industrial Classification categorizes health care as generally consisting of hospital activities, medical and dental practice activities, and "other human health activities". The last class involves activities of, or under the supervision of, nurses, midwives, physiotherapists, scientific or diagnostic laboratories, pathology clinics, residential health facilities, patient advocates, or other allied health professions, e.g. in the field of optometry, hydrotherapy, medical massage, yoga therapy, music therapy, occupational therapy, speech therapy, chiropody, homeopathy, chiropractics, acupuncture, etc.

In addition, according to industry and market classifications, such as the Global Industry Classification Standard and the Industry Classification Benchmark, health care includes many categories of medical equipment, instruments and services as well as biotechnology, diagnostic laboratories and substances, and drug manufacturing and delivery.

For example, pharmaceuticals and other medical devices are the leading high technology exports of Europe and the United States. The United States dominates the biopharmaceutical field, accounting for three-quarters of the world's biotechnology revenues.

Health Care Research

The quantity and quality of many health care interventions are improved through the results of science, such as advanced through the medical model of health which focuses on the eradication of illness through diagnosis and effective treatment. Many important advances have been made through health research, including biomedical research and pharmaceutical research, which form the basis for evidence-based medicine and evidence-based practice in health care delivery.

For example, in terms of pharmaceutical research and development spending, Europe spends a little less than the United States (€22.50bn compared to €27.05bn in 2006). The United States accounts for 80% of the world's research and development spending in biotechnology.

In addition, the results of health services research can lead to greater efficiency and equitable delivery of health care interventions, as advanced through the social model of health and disability, which emphasizes the societal changes that can be made to make population healthier. Results from health services research often form the basis of evidence-based policy in health care systems. Health services research is also aided by initiatives in the field of AI for the development of systems of health assessment that are clinically useful, timely, sensitive to change, culturally sensitive, low burden, low cost, involving for the patient and built into standard procedures.

Health Care Financing

There are generally five primary methods of funding health care systems:

1. general taxation to the state, county or municipality

2. social health insurance

3. voluntary or private health insurance

4. out-of-pocket payments

5. donations to health charities

In most countries, the financing of health care services features a mix of all five models, but the exact distribution varies across countries and over time within countries. In all countries and jurisdictions, there are many topics in the politics and evidence that can influence the decision of a government, private sector business or other group to adopt a specific health policy regarding the financing structure.

For example, social health insurance is where a nation's entire population is eligible for health care coverage, and this coverage and the services provided are regulated. In almost every jurisdiction with a government-funded health care system, a parallel private, and usually for-profit, system is allowed to operate. This is sometimes referred to as two-tier health care or universal health care.

For example, in Poland, the costs of health services borne by the National Health Fund (financed by all citizens that pay health insurance contributions) in 2012 amounted to 60.8 billion PLN (approximately 20 billion USD). The right to health services in Poland is granted to 99.9% of the population (also registered unemployed persons and their spouses).

Health Care Administration and Regulation

The management and administration of health care is another sector vital to the delivery of health care services. In particular, the practice of health professionals and operation of health care institutions is typically regulated by national or state/provincial authorities through appropriate regulatory bodies for purposes of quality assurance. Most countries have credentialing staff in regulatory boards or health departments who document the certification or licensing of health workers and their work history.

Health Information Technology

Health information technology (HIT) is "the application of information processing involving both computer hardware and software that deals with the storage, retrieval, sharing, and use of health care information, data, and knowledge for communication and decision making." Technology is a broad concept that deals with a species' usage and knowledge of tools and crafts, and how it affects a species' ability to control and adapt to its environment. However, a strict definition is elusive; "technology" can refer to material objects of use to humanity, such as machines, hardware or utensils, but can also encompass broader themes, including systems, methods of organization, and techniques. For HIT, technology represents computers and communications attributes that can be networked to build systems for moving health information. Informatics is yet another integral aspect of HIT.

Health information technology can be divided into further components like Electronic Health Record (EHR), Electronic Medical Record (EMR), Personal Health Record (PHR), Practice Management System (PMS), Health Information Exchange (HIE) and many more. There are multiple purposes for the use of HIT within the health care industry. Further, the use of HIT is expected to improve the quality of health care, reduce medical errors, improve the health care service efficiency and reduce health care costs.

In Vitro Fertilisation

In vitro fertilisation (or fertilization; IVF) is a process by which an egg is fertilised by sperm outside the body: *in vitro* ("in glass"). The process involves monitoring and stimulating a woman's ovulatory process, removing an ovum or ova (egg or eggs) from the woman's ovaries and letting sperm fertilise them in a liquid in a laboratory. The fertilised egg (zygote) is cultured for 2–6 days in a growth medium and is then implanted in the same or another woman's uterus, with the intention of establishing a successful pregnancy.

IVF techniques can be used in different types of situations. It is a technique of assisted reproductive technology for treatment of infertility. IVF techniques are also employed in gestational surrogacy, in which case the fertilised egg is implanted into a surrogate's uterus, and the resulting child is genetically unrelated to the surrogate. In some situations, donated eggs or sperms may be used. Some countries ban or otherwise regulate the availability of IVF treatment, giving rise to fertility tourism. Restrictions on availability of IVF include costs and age to carry a healthy pregnancy to term. Due to the costs of the procedure, IVF is mostly attempted only after less expensive options have failed.

The first successful birth of a "test tube baby", Louise Brown, occurred in 1978. Louise Brown was born as a result of natural cycle IVF where no stimulation was made. Robert G. Edwards, the physiologist who developed the treatment, was awarded the Nobel Prize in Physiology or Medicine in 2010. With egg donation and IVF, women who are past their reproductive years or menopause can still become pregnant. Adriana Iliescu held the record as the oldest woman to give birth using IVF and donated egg, when she gave birth in 2004 at the age of 66, a record passed in 2006. After the IVF treatment many couples are able to get pregnant without any fertility treatments.

Terminology

The term *in vitro*, from the Latin meaning *in glass*, is used, because early biological experiments involving cultivation of tissues outside the living organism from which they came, were carried out in glass containers such as *beakers, test tubes, or petri dishes*.

Today, the scientific term *in vitro* is used to refer to any biological procedure that is performed outside the organism in which it would normally have occurred, to distinguish it from an *in vivo* procedure, where the tissue remains inside the living organism within which it is normally found. A colloquial term for babies conceived as the result of IVF, "test tube babies", refers to the tube-shaped containers of glass or plastic resin, called *test tubes,* that are commonly used in chemistry labs and biology labs. However, *in vitro* fertilisation is usually performed in the shallower containers called Petri dishes. One IVF method, autologous endometrial coculture, is actually performed on organic material, but is still considered *in vitro*.

Medical Uses

IVF may be used to overcome female infertility where it is due to problems with the fallopian tubes, making fertilisation *in vivo* difficult. It can also assist in male infertility, in those cases where there is a defect in sperm quality; in such situations intracytoplasmic sperm injection (ICSI) may be used, where a sperm cell is injected directly into the egg cell. This is used when sperm has difficulty penetrating the egg, and in these cases the partner's or a donor's sperm may be used. ICSI is also used when sperm numbers are very low. When indicated, the use of ICSI has been found to increase the success rates of IVF.

According to the British NICE guidelines, IVF treatment is appropriate in cases of unexplained infertility for women that have not conceived after 2 years of regular unprotected sexual intercourse. This rule does not apply to all countries.

IVF is also considered suitable in cases where any of its expansions is of interest, that is, a procedure that is usually not necessary for the IVF procedure itself, but would be virtually impossible or technically difficult to perform without concomitantly performing methods of IVF. Such expansions include preimplantation genetic diagnosis (PGD) to rule out presence of genetic disorders, as well as egg donation or surrogacy where the woman providing the egg isn't the same who will carry the pregnancy to term. Further details in the Expansions-section below.

Success Rates

IVF success rates are the percentage of all IVF procedures which result in a favorable outcome. Depending on the type of calculation used, this outcome may represent the number of confirmed pregnancies, called the pregnancy rate, or the number of live births, called the live birth rate. The success rate depends on variable factors such as maternal age, cause of infertility, embryo status, reproductive history and lifestyle factors.

Maternal age: Younger candidates of IVF are more likely to get pregnant. Women older than 41 are more likely to get pregnant with a donor egg.

Reproductive history: Women who have been previously pregnant are in many cases more successful with IVF treatments then those who have never been pregnant.

Due to advances in reproductive technology, IVF success rates are substantially higher today than they were just a few years ago.

Live Birth Rate

The live birth rate is the percentage of all IVF cycles that lead to a live birth. This rate does not include miscarriage or stillbirth and multiple-order births such as twins and triplets are counted as one pregnancy. A 2012 summary compiled by the Society for Reproductive Medicine which reports the average IVF success rates in the United States per age group using non-donor eggs compiled the following data:

	<35	35-37	38-40	41-42	>42
Pregnancy rate	46.7	37.8	29.7	19.8	8.6
Live birth rate	40.7	31.3	22.2	11.8	3.9

In 2006, Canadian clinics reported a live birth rate of 27%. Birth rates in younger patients were slightly higher, with a success rate of 35.3% for those 21 and younger, the youngest group evaluated. Success rates for older patients were also lower and decrease with age, with 37-year-olds at 27.4% and no live births for those older than 48, the oldest group evaluated. Some clinics exceeded these rates, but it is impossible to determine if that is due to superior technique or patient selection, because it is possible to artificially increase success rates by refusing to accept the most difficult patients or by steering them into oocyte donation cycles (which are compiled separately). Further, pregnancy rates can be increased by the placement of several embryos at the risk of increasing the chance for multiples.

The live birth rates using donor eggs are also given by the SART and include all age groups using either fresh or thawed eggs.

	Fresh donor egg embryos	Thawed donor egg embryos
Live birth rate	55.1	33.8

Because not each IVF cycle that is started will lead to oocyte retrieval or embryo transfer, reports of live birth rates need to specify the denominator, namely IVF cycles started, IVF retrievals, or embryo transfers. The Society for Assisted Reproductive Technology (SART) summarised 2008-9 success rates for US clinics for fresh embryo cycles that did not involve donor eggs and gave live birth rates by the age of the prospective mother, with a peak at 41.3% per cycle started and 47.3% per embryo transfer for patients under 35 years of age.

IVF attempts in multiple cycles result in increased cumulative live birth rates. Depending on the demographic group, one study reported 45% to 53% for three attempts, and 51% to 71% to 80% for six attempts.

Pregnancy Rate

Pregnancy rate may be defined in various ways. In the United States, the pregnancy rate used by the Society for Assisted Reproductive Technology and the Centers for Disease Control (and appearing in the table in the Success Rates section above) are based on fetal heart motion observed in ultrasound examinations.

The 2009 summary compiled by the Society for Reproductive Medicine included the following data for the United States:

	<35	35-37	38-40	41-42
Pregnancy rate	47.6	38.9	30.1	20.5

In 2006, Canadian clinics reported an average pregnancy rate of 35%. A French study estimated that 66% of patients starting IVF treatment finally succeed in having a child (40% during the IVF treatment at the center and 26% after IVF discontinuation). Achievement of having a child after IVF discontinuation was mainly due to adoption (46%) or spontaneous pregnancy (42%).

Predictors of Success

The main potential factors that influence pregnancy (and live birth) rates in IVF have been suggested to be maternal age, duration of infertility or subfertility, bFSH and number of oocytes, all reflecting ovarian function. Optimal woman's age is 23–39 years at time of treatment.

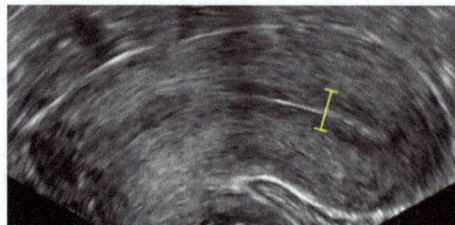

A triple-line endometrium is associated with better IVF outcomes.

Biomarkers that affect the pregnancy chances of IVF include:

- Antral follicle count, with higher count giving higher success rates.

- Anti-Müllerian hormone levels, with higher levels indicating higher chances of pregnancy, as well as of live birth after IVF, even after adjusting for age.

- Factors of semen quality for the sperm provider.

- Level of DNA fragmentation as measured e.g. by Comet assay, advanced maternal age and semen quality.

- Women with ovary-specific FMR1 genotypes including *het-norm/low* have significantly decreased pregnancy chances in IVF.

- Progesterone elevation (PE) on the day of induction of final maturation is associated with lower pregnancy rates in IVF cycles in women undergoing ovarian stimulation using GnRH analogues and gonadotrophins. At this time, compared to a progesterone level below 0.8 ng/ml, a level between 0.8 and 1.1 ng/ml confers an odds ratio of pregnancy of approximately 0.8, and a level between 1.2 and 3.0 ng/ml confers an odds ratio of pregnancy of between 0.6 and 0.7. On the other hand, progesterone elevation does not seem to confer a decreased chance of pregnancy in frozen–thawed cycles and cycles with egg donation.

- Characteristics of cells from the cumulus oophorus and the membrana granulosa, which are easily aspirated during oocyte retrieval. These cells are closely associated with the oocyte and share the same microenvironment, and the rate of expression of certain genes in such cells are associated with higher or lower pregnancy rate.

- An endometrial thickness (EMT) of less than 7 mm decreases the pregnancy rate by an odds ratio of approximately 0.4 compared to an EMT of over 7 mm. However, such low thickness rarely occurs, and any routine use of this parameter is regarded as not justified.

Other determinants of outcome of IVF include:

- Tobacco smoking reduces the chances of IVF producing a live birth by 34% and increases the risk of an IVF pregnancy miscarrying by 30%.

- A body mass index (BMI) over 27 causes a 33% decrease in likelihood to have a live birth after the first cycle of IVF, compared to those with a BMI between 20 and 27. Also, pregnant women who are obese have higher rates of miscarriage, gestational diabetes, hypertension, thromboembolism and problems during delivery, as well as leading to an increased risk of fetal congenital abnormality. Ideal body mass index is 19–30.

- Salpingectomy or laparoscopic tubal occlusion before IVF treatment increases chances for women with hydrosalpinges

- Success with previous pregnancy and/or live birth increases chances

- Low alcohol/caffeine intake increases success rate

- The number of embryos transferred in the treatment cycle.

- Embryo quality

- Some studies also suggest the autoimmune disease may also play a role in decreasing IVF success rates by interfering with proper implantation of the embryo after transfer.

Aspirin is sometimes prescribed to women for the purpose of increasing the chances of conception by IVF, but there is insufficient evidence to show that it actually works.

A 2013 review and metaanalysis of randomised controlled trials of acupuncture as an adjuvant therapy in IVF found no overall benefit, and concluded that an apparent benefit detected in a subset of published trials where the control group (those not using acupuncture) experienced a lower than average rate of pregnancy requires further study, due to the possibility of publication bias and other factors.

A Cochrane review came to the result that endometrial injury performed in the month prior to ovarian hyperstimulation appeared to increase both the live birth rate and clinical pregnancy rate in IVF compared with no endometrial injury. However, there was a lack of data reported on the rates of adverse outcomes such as miscarriage, multiple pregnancy, pain and/or bleeding.

For women, intake of antioxidants (such as N-acetyl-cysteine, melatonin, vitamin A, vitamin C, vitamin E, folic acid, myo-inositol, zinc or selenium) has not been associated with a significantly increased live birth rate or clinical pregnancy rate in IVF according to Cochrane reviews. On the other hand, oral antioxidants given to the men in couples with male factor or unexplained subfertility resulted in significantly higher live birth rate in IVF.

A Cochrane review in 2013 came to the result that there is no evidence identified regarding the effect of pre-conception lifestyle advice on the chance of a live birth outcome.

Complications

Multiple Births

The major complication of IVF is the risk of multiple births. This is directly related to the practice of transferring multiple embryos at embryo transfer. Multiple births are related to increased risk of pregnancy loss, obstetrical complications, prematurity, and neonatal morbidity with the potential for long term damage. Strict limits on the number of embryos that may be transferred have been enacted in some countries (e.g. Britain, Belgium) to reduce the risk of high-order multiples (triplets or more), but are not universally followed or accepted. Spontaneous splitting of embryos in the womb after transfer can occur, but this is rare and would lead to identical twins. A double blind, randomised study followed IVF pregnancies that resulted in 73 infants (33 boys and 40 girls) and reported that 8.7% of singleton infants and 54.2% of twins had a birth weight of < 2,500 grams (5.5 lb).

Recent evidence also suggest that singleton offspring after IVF is at higher risk for lower birth weight for unknown reasons.

Spread of Infectious Disease

By sperm washing, the risk that a chronic disease in the male providing the sperm would infect the female or offspring can be brought to negligible levels.

In males with hepatitis B, The Practice Committee of the American Society for Reproductive Medicine advises that sperm washing is not necessary in IVF to prevent transmission, unless the female partner has not been effectively vaccinated. In females with hepatitis B, the risk of vertical transmission during IVF is no different from the risk in spontaneous conception. However, there is not enough evidence to say that ICSI procedures are safe in females with hepatitis B in regard to vertical transmission to the offspring.

Regarding potential spread of HIV/AIDS, Japan's government prohibited the use of *in vitro* fertilisation procedures for couples in which both partners are infected with HIV. Despite the fact that the ethics committees previously allowed the Ogikubo, Tokyo Hospital, located in Tokyo, to use *in vitro* fertilisation for couples with HIV, the Ministry of Health, Labour and Welfare of Japan decided to block the practice. Hideji Hanabusa, the vice president of the Ogikubo Hospital, states that together with his colleagues, he managed to develop a method through which scientists are able to remove HIV from sperm.

Other Risks to the Egg Provider/Retriever

A risk of ovarian stimulation is the development of ovarian hyperstimulation syndrome, particularly if hCG is used for inducing final oocyte maturation. This results in swollen, painful ovaries. It occurs in 30% of patients. Mild cases can be treated with over the counter medications and cases can be resolved in the absence of pregnancy. In moderate cases, ovaries swell and fluid accumulated in the abdominal cavities and may have symptoms of heartburn, gas, nausea or loss of appetite. In severe cases patients have sudden excess abdominal pain, nausea, vomiting and will result in hospitalisation.

During egg retrieval, there's a small chance of bleeding, infection, and damage to surrounding structures like bowel and bladder (transvaginal ultrasound aspiration) as well as difficulty in breathing, chest infection, allergic reactions to medication, or nerve damage (laproscopy).

Ectopic pregnancy may also occur if a fertilised egg develops outside the uterus, usually in the fallopian tubes and requires immediate destruction of the foetus.

IVF does not seem to be associated with an elevated risk of cervical cancer, nor with ovarian cancer or endometrial cancer when neutralising the confounder of infertility itself. Nor does it seem to impart any increased risk for breast cancer.

Regardless of pregnancy result, IVF treatment is usually stressful for patients. Neuroticism and the use of escapist coping strategies are associated with a higher degree of distress, while the presence social support has a relieving effect. A negative pregnancy test after IVF is associated with an increased risk for depression in women, but not with any increased risk of developing anxiety disorders. Pregnancy test results do not seem to be a risk factor for depression or anxiety among men.

Birth Defects

A review in 2013 came to the result that infants resulting from IVF (with or without ICSI) have a relative risk of birth defects of 1.32 (95% confidence interval 1.24–1.42) compared to naturally conceived infants. In 2008, an analysis of the data of the National Birth Defects Study in the US found that certain birth defects were significantly more common in infants conceived through IVF, notably septal heart defects, cleft lip with or without cleft palate, esophageal atresia, and anorectal atresia; the mechanism of causality is unclear. However, in a population-wide cohort study of 308,974 births (with 6163 using assisted reproductive technology and following children from birth to age five) researchers found: "The increased risk of birth defects associated with IVF was no longer significant after adjustment for parental factors." Parental factors included known independent risks for birth defects such as maternal age, smoking status, etc. Multivariate correction did not remove the significance of the association of birth defects and ICSI (corrected odds ratio 1.57), although the authors speculate that underlying male infertility factors (which would be associated with the use of ICSI) may contribute to this observation and were not able to correct for these confounders. The authors also found that a history of infertility elevated risk itself in the absence of any treatment (odds ratio 1.29), consistent with a Danish national registry study and "...implicates patient factors in this increased risk." The authors of the Danish national registry study speculate: "...our results suggest that the reported increased prevalence of congenital malformations seen in singletons born after assisted reproductive technology is partly due to the underlying infertility or its determinants."

Risk in singleton pregnancies resulting from IVF (with or without ICSI)		
Condition	**Relative risk**	**95% confidence interval**
Beckwith–Wiedemann syndrome	3-4	
congenital anomalies	1.67	1.33–2.09
ante-partum haemorrhage	2.49	2.30–2.69
hypertensive disorders of pregnancy	1.49	1.39–1.59
preterm rupture of membranes	1.16	1.07–1.26
Caesarean section	1.56	1.51–1.60
gestational diabetes	1.48	1.33–1.66
induction of labour	1.18	1.10–1.28
small for gestational age	1.39	1.27–1.53
preterm birth	1.54	1.47–1.62
low birthweight	1.65	1.56–1.75
perinatal mortality	1.87	1.48–2.37

Other Risks to the Offspring

If the underlying infertility is related to abnormalities in spermatogenesis, it is plausible, but too early to examine that male offspring are at higher risk for sperm abnormalities.

IVF does not seem to confer any risks regarding cognitive development, school performance, social functioning and behaviour. Also, IVF infants are known to be as securely attached to their parents as those who were naturally conceived, and IVF adolescents are as well-adjusted as those who have been naturally conceived.

Limited long-term follow-up data suggest that IVF may be associated with an increased incidence of hypertension, impaired fasting glucose, increase in total body fat composition, advancement of bone age, subclinical thyroid disorder, early adulthood clinical depression and binge drinking in the offspring. It is not known, however, whether these potential associations are caused by the IVF procedure in itself, by adverse obstetric outcomes associated with IVF, by the genetic origin of the children or by yet unknown IVF-associated causes. Increases in embryo manipulation during IVF result in more deviant fetal growth curves, but birth weight does not seem to be a reliable marker of fetal stress.

IVF, including ICSI, is associated with an increased risk of imprinting disorders (including Prader-Willi syndrome and Angelman syndrome), with an odds ratio of 3.7 (95% confidence interval 1.4 to 9.7).

An IVF-associated incidence of cerebral palsy and neurodevelopmental delay are believed to be related to the confounders of prematurity and low birthweight. Similarly, an IVF-associated incidence of autism and attention-deficit disorder are believed to be related to confounders of maternal and obstetric factors.

Overall, IVF does not cause an increased risk of childhood cancer. Studies have shown a decrease in the risk of certain cancers and an increased risks of certain others including retinoblastoma hepatoblastoma and rhabdomyosarcoma.

Method

Theoretically, *in vitro* fertilisation could be performed by collecting the contents from a woman's fallopian tubes or uterus after natural ovulation, mixing it with sperm, and reinserting the fertilised ova into the uterus. However, without additional techniques, the chances of pregnancy would be extremely small. The additional techniques that are routinely used in IVF include ovarian hyperstimulation to generate multiple eggs or ultrasound-guided transvaginal oocyte retrieval directly from the ovaries; after which the ova and sperm are prepared, as well as culture and selection of resultant embryos before embryo transfer into a uterus.

Ovarian Hyperstimulation

Ovarian hyperstimulation is the stimulation to induce development of multiple follicles of the ovaries. It should start with response prediction by e.g. age, antral follicle count and level of anti-Müllerian hormone. The resulting prediction of e.g. poor or hyper-response to ovarian hyperstimulation determines the protocol and dosage for ovarian hyperstimulation.

Ovarian hyperstimulation also includes suppression of spontaneous ovulation, for which two main methods are available: Using a (usually longer) GnRH agonist protocol or a (usually shorter) GnRH antagonist protocol. In a standard long GnRH agonist protocol the day when hyperstimulation treatment is started and the expected day of later oocyte retrieval can be chosen to conform to personal choice, while in a GnRH antagonist protocol it must be adapted to the spontaneous onset of the previous menstruation. On the other hand, the GnRH antagonist protocol has a lower risk of ovarian hyperstimulation syndrome (OHSS), which is a life-threatening complication.

For the ovarian hyperstimulation in itself, injectable gonadotropins (usually FSH analogues) are generally used under close monitoring. Such monitoring frequently checks the estradiol level and, by means of gynecologic ultrasonography, follicular growth. Typically approximately 10 days of injections will be necessary.

Natural IVF

There are several methods termed *natural cycle IVF*:

- IVF using no drugs for ovarian hyperstimulation, while drugs for ovulation suppression may still be used.

- IVF using ovarian hyperstimulation, including gonadotropins, but with a GnRH antagonist protocol so that the cycle initiates from natural mechanisms.

- Frozen embryo transfer; IVF using ovarian hyperstimulation, followed by embryo cryopreservation, followed by embryo transfer in a later, natural, cycle.

IVF using no drugs for ovarian hyperstimulation was the method for the conception of Louise Brown. This method can be successfully used when women want to avoid taking ovarian stimulating drugs with its associated side-effects. HFEA has estimated the live birth rate to be approximately 1.3% per IVF cycle using no hyperstimulation drugs for women aged between 40–42.

Mild IVF is a method where a small dose of ovarian stimulating drugs are used for a short duration during a woman's natural cycle aimed at producing 2–7 eggs and creating healthy embryos. This method appears to be an advance in the field to reduce complications and side-effects for women and it is aimed at quality, and not quantity of eggs and embryos. One study comparing a mild treatment (mild ovarian stimulation with GnRH antagonist co-treatment combined with single embryo transfer) to a standard treatment (stimulation with a GnRH agonist long-protocol and transfer of two embryos) came to the result that the proportions of cumulative pregnancies that resulted in term live birth after 1 year were 43.4% with mild treatment and 44.7% with standard treatment. Mild IVF can be cheaper than conventional IVF and with a significantly reduced risk of multiple gestation and OHSS.

Final Maturation Induction

When the ovarian follicles have reached a certain degree of development, induction of final oocyte maturation is performed, generally by an injection of human chorionic gonadotropin (hCG). Commonly, this is known as the "trigger shot." hCG acts as an analogue of luteinising hormone, and ovulation would occur between 38 and 40 hours after a single HCG injection, but the egg retrieval is performed at a time usually between 34 and 36 hours after hCG injection, that is, just prior to when the follicles would rupture. This avails for scheduling the egg retrieval procedure at a time where the eggs are fully mature. HCG injection confers a risk of ovarian hyperstimulation syndrome. Using a GnRH agonist instead of hCG eliminates the risk of ovarian hyperstimulation syndrome, but with a delivery rate of approximately 6% less than with hCG.

Egg Retrieval

The eggs are retrieved from the patient using a transvaginal technique called transvaginal oocyte retrieval, involving an ultrasound-guided needle piercing the vaginal wall to reach the ovaries. Through this needle follicles can be aspirated, and the follicular fluid is passed to an embryologist to identify ova. It is common to remove between ten and thirty eggs. The retrieval procedure usually takes between 20 and 40 minutes, depending on the number of mature follicles, and is usually done under conscious sedation or general anaesthesia.

Egg and Sperm Preparation

In the laboratory, the identified eggs are stripped of surrounding cells and prepared for fertilisation. An oocyte selection may be performed prior to fertilisation to select eggs with optimal chances of successful pregnancy. In the meantime, semen is prepared for fertilisation by removing inactive cells and seminal fluid in a process called sperm washing. If semen is being provided by a sperm donor, it will usually have been prepared for treatment before being frozen and quarantined, and it will be thawed ready for use.

Co-incubation

The sperm and the egg are incubated together at a ratio of about 75,000:1 in a culture media in order for the actual fertilisation to take place. A review in 2013 came to the result that a duration of this co-incubation of about 1 to 4 hours results in significantly higher pregnancy rates than 16 to 24 hours. In most cases, the egg will be fertilised during co-incubation and will show two pronuclei. In certain situations, such as low sperm count or motility, a single sperm may be injected directly into the egg using intracytoplasmic sperm injection (ICSI). The fertilised egg is passed to a special growth medium and left for about 48 hours until the egg consists of six to eight cells.

In gamete intrafallopian transfer, eggs are removed from the woman and placed in one of the fallopian tubes, along with the man's sperm. This allows fertilisation to take place inside the woman's body. Therefore, this variation is actually an *in vivo* fertilisation, not an *in vitro* fertilisation.

Embryo Culture

The main durations of embryo culture are until cleavage stage (day two to four after co-incubation) or the blastocyst stage (day five or six after co-incubation). Embryo culture until the blastocyst stage confers a significant increase in live birth rate per embryo transfer, but also confers a decreased number of embryos available for transfer and embryo cryopreservation, so the cumulative clinical pregnancy rates are increased with cleavage stage transfer. Transfer day two instead of day three after fertilisation has no differences in live birth rate. There are significantly higher odds of preterm birth (odds ratio 1.3) and congenital anomalies (odds ratio 1.3) among births having from embryos cultured until the blastocyst stage compared with cleavage stage.

Embryo Selection

Laboratories have developed grading methods to judge oocyte and embryo quality. In order to optimise pregnancy rates, there is significant evidence that a morphological scoring system is the best strategy for the selection of embryos. Since 2009 where the first time-lapse microscopy system for IVF was approved for clinical use, morphokinetic scoring systems has shown to improve to pregnancy rates further. However, when all different types of time-lapse embryo imaging devices, with or without morphokinetic scoring systems, are compared against conventional embryo assessment for IVF, there is insufficient evidence of a difference in live-birth, pregnancy, stillbirth or miscarriage to choose between them.

Embryo Transfer

Embryos are failed by the embryologist based on the amount of cells, evenness of growth and degree of fragmentation. The number to be transferred depends on the number available, the age of the woman and other health and diagnostic factors. In countries such as Canada, the UK, Australia and New Zealand, a maximum of two embryos are transferred except in unusual circumstances. In the UK and according to HFEA regulations, a woman over 40 may have up to three embryos transferred, whereas in the USA, younger women may have many embryos transferred based on individual fertility diagnosis. Most clinics and country regulatory bodies seek to minimise the risk of pregnancies carrying multiples, as it is not uncommon for more implantations to take than desired. The embryos judged to be the "best" are transferred to the patient's uterus through a thin, plastic catheter, which goes through her vagina and cervix. Several embryos may be passed into the uterus to improve chances of implantation and pregnancy.

Adjunctive Medication

Luteal support is the administration of medication, generally progesterone, progestins or GnRH agonists, to increase the success rate of implantation and early embryogenesis, thereby complementing and/or supporting the function of the corpus luteum. The live birth rate is significantly higher with progesterone for luteal support in IVF cycles with or without intracytoplasmic sperm injection (ICSI).Co-treatment with GnRH agonists further improves outcomes, by a live birth rate RD of +16% (95% confidence interval +10 to +22%).

On the other hand, growth hormone or aspirin as adjunctive medication in IVF have no evidence of overall benefit.

Expansions

There are various expansions or additional techniques that can be applied in IVF, which are usually not necessary for the IVF procedure itself, but would be virtually impossible or technically difficult to perform without concomitantly performing methods of IVF.

Preimplantation Genetic Screening or Diagnosis

Preimplantation genetic screening (PGS) or preimplantation genetic diagnosis (PGD) has been suggested to be able to be used in IVF to select an embryo that appears to have the greatest chances for successful pregnancy. However, a systematic review and meta-analysis of existing randomised controlled trials came to the result that there is no evidence of a beneficial effect of PGS with cleavage-stage biopsy as measured by live birth rate. On the contrary, for women of advanced maternal age, PGS with cleavage-stage biopsy significantly lowers the live birth rate. Technical drawbacks, such as the invasiveness of the biopsy, and non-representative samples because of mosaicism are the major underlying factors for inefficacy of PGS.

Still, as an expansion of IVF, patients who can benefit from PGS/PGD include:

- Couples who have a family history of inherited disease

- Couples who want to use gender selection to prevent a gender-linked disease

- Couples who already have a child with an incurable disease and need compatible cells from a second healthy child to cure the first, resulting in a "saviour sibling" that matches the sick child in HLA type.

PGS screens for numeral chromosomal abnormalities while PGD diagnosis the specific molecular defect of the inherited disease. In both PGS and PGD, individual cells from a pre-embryo, or preferably trophectoderm cells biopsied from a blastocyst, are analyzed during the IVF process. Before the transfer of a pre-embryo back to a woman's uterus, one or two cells are removed from the pre-embryos (8-cell stage), or preferably from a blastocyst. These cells are then evaluated for normality. Typically within one to two days, following completion of the evaluation, only the normal pre-embryos are transferred back to the woman's uterus. Alternatively, a blastocyst can be cryopreserved via vitrification and transferred at a later date to the uterus. In addition, PGS can significantly reduce the risk of multiple pregnancies because fewer embryos, ideally just one, are needed for implantation.

Cryopreservation

Cryopreservation can be performed as oocyte cryopreservation before fertilisation, or as embryo cryopreservation after fertilisation.

The Rand Consulting Group has estimated there to be 400,000 frozen embryos in the United States. The advantage is that patients who fail to conceive may become pregnant using such embryos without having to go through a full IVF cycle. Or, if pregnancy occurred, they could return later for another pregnancy. Spare oocytes or embryos resulting from fertility treatments may be used for oocyte donation or embryo donation to another woman or couple, and embryos may be created, frozen and stored specifically for transfer and donation by using donor eggs and sperm. Also, oocyte cryopreservation can be used for women who are likely to lose their ovarian reserve due to undergoing chemotherapy.

The outcome from using cryopreserved embryos has uniformly been positive with no increase in birth defects or development abnormalities.

Other Expansions

- Intracytoplasmic sperm injection (*ICSI*) is where a single sperm is injected directly into an egg. Its main usage as an expansion of IVF is to overcome male infertility problems, although it may also be used where eggs cannot easily be penetrated by sperm, and occasionally in conjunction with sperm donation. It can be used in teratozoospermia, since once the egg is fertilised abnormal sperm morphology does not appear to influence blastocyst development or blastocyst morphology.

- Additional methods of embryo profiling. For example, methods are emerging in making comprehensive analyses of up to entire genomes, transcriptomes, proteomes and metabolomes which may be used to score embryos by comparing the patterns with ones that have previously been found among embryos in successful versus unsuccessful pregnancies.

- Assisted zona hatching (AZH) can be performed shortly before the embryo is transferred to the uterus. A small opening is made in the outer layer surrounding the egg in order to help the embryo hatch out and aid in the implantation process of the growing embryo.

- In egg donation and embryo donation, the resultant embryo after fertilisation is inserted in another woman than the one providing the eggs. These are resources for women with no eggs due to surgery, chemotherapy, or genetic causes; or with poor egg quality, previously unsuccessful IVF cycles or advanced maternal age. In the egg donor process, eggs are retrieved from a donor's ovaries, fertilised in the laboratory with the sperm from the recipient's partner, and the resulting healthy embryos are returned to the recipient's uterus.

- In oocyte selection, the oocytes with optimal chances of live birth can be chosen. It can also be used as a means of preimplantation genetic screening.

- Embryo splitting can be used for twinning to increase the number of available embryos.

Cytoplasmic Transfer

Cytoplasmic transfer was created to aid women who experience infertility due to deficient or damaged mitochondria, contained within an egg's cytoplasm. Deficient mitochondria can lead to recurrent implantation failure, high levels of embryo fragmentation and overall poor embryo development. The incidence of compromised mitochondria increases with advanced maternal age, thought to occur near the age of thirty-five. Consequently, it has been found advantageous for young women to donate cytoplasm to older women, creating rejuvenated eggs.

Though cytoplasmic transfer does not involve the transfer of nuclear DNA, there may still be a small amount of mitochondrial DNA present from the donor. Children conceived through this process occasionally test positive for genetic material from three parents. There is much concern associated with the potential transfer of mitochondrial DNA and its unknown interaction with the foreign DNA of the recipient egg. There is also no data as to the health of maturing children conceived through cytoplasmic transfer, the first successful birth having occurred in 1997 at Saint Barnabas Medical Center in Livingston, New Jersey as a result of procedure performed by The Institute for Reproductive Medicine and Science. The procedure is not currently approved for general use in any country aside from the United Kingdom, which legalised it in October 2015.

Leftover Embryos or Eggs

There may be leftover embryos or eggs from IVF procedures if the woman for whom they were originally created has successfully carried one or more pregnancies to term. With the woman's or couple's permission, these may be donated to help other women or couples as a means of third party reproduction.

In embryo donation, these extra embryos are given to other couples or women for transfer with the goal of producing a successful pregnancy. The resulting child is considered the child of the woman who carries it and gives birth, and not the child of the donor, the same as occurs with egg donation or sperm donation.

Typically, genetic parents donate the eggs to a fertility clinic orwhere they are preserved by oocyte cryopreservation or embryo cryopreservation until a carrier is found for them. Typically the process of matching the embryo(s) with the prospective parents is conducted by the agency itself, at which time the clinic transfers ownership of the embryos to the prospective parents.

In the United States, women seeking to be an embryo recipient undergo infectious disease screening required by the U.S. Food and Drug Administration (FDA), and reproductive tests to determine the best placement location and cycle timing before the actual Embryo Transfer occurs. The amount of screening the embryo has already undergone is largely dependent on the genetic parents' own IVF clinic and process. The embryo recipient may elect to have her own embryologist conduct further testing.

Alternatives to donating unused embryos are destroying them (or having them implanted at a time where pregnancy is very unlikely), keeping them frozen indefinitely, or donating them for use in research (which results in their unviability). Individual moral views on disposing leftover embryos may depend on personal views on the beginning of human personhood and definition and/or value of potential future persons and on the value that is given to fundamental research questions. Some people believe donation of leftover embryos for research is a good alternative to discarding the embryos when patients receive proper, honest and clear information about the research project, the procedures and the scientific values.

History

In 1977, Steptoe and Edwards successfully carried out a pioneering conception which resulted in the birth of the world's first baby to be conceived by IVF, Louise Brown on 25 July 1978, in Oldham General Hospital, Greater Manchester, UK.

The second successful birth of a test tube baby occurred in India just 67 days after Louise Brown was born. The girl, named Durga conceived *in vitro* using the methods of Subhash Mukhopadhyay, a physician and researcher from Kolkata.

Ethics

Mix-ups

In some cases, laboratory mix-ups (misidentified gametes, transfer of wrong embryos) have occurred, leading to legal action against the IVF provider and complex paternity suits. An example is the case of a woman in California who received the embryo of another couple and was notified of this mistake after the birth of her son. This has led to many authorities and individual clinics implementing procedures to minimise the risk of such mix-ups. The HFEA, for example, requires clinics to use a double witnessing system, the identity of specimens is checked by two people at each point at which specimens are transferred. Alternatively, technological solutions are gaining favour, to reduce the manpower cost of manual double witnessing, and to further reduce risks with uniquely numbered RFID tags which can be identified by readers connected to a computer. The computer tracks specimens throughout the process and alerts the embryologist if non-matching specimens are identified. Although the use of RFID tracking has expanded in the USA, it is still not widely adopted. However, In other cases there has been not mix-up of embryos or gametes, but the intentional use of embryos of another couple or gamete donor, without informed consent of parents, both: receptors or donors. Some of these cases are taking a legal and judicial course.

Preimplantation Genetic Diagnosis or Screening

Another concern is that people will screen in or out for particular traits, using preimplantation genetic diagnosis (PGD) or preimplantation genetic screening. For example, a deaf British couple, Tom and Paula Lichy, have petitioned to create a deaf baby using IVF. Some medical ethicists have been very critical of this approach. Jacob M. Appel wrote that "intentionally culling out blind or deaf embryos might prevent considerable future suffering, while a policy that allowed deaf or blind parents to select *for* such traits intentionally would be far more troublesome."

This concept of decisively altering genes has coined the concept of the Designer Baby.

Currently, PGD can alter some physical and health attributes, and projections for the future power of PGD in its ability to create the ideal human has raised many ethical issues. Projections for societal repercussions include changing the realm athletics, creating human weapons, and exchanging autonomy over one's life course for predesignation. Also, with a limited view of the future, it is difficult to alter a human's genetic makeup without knowing full repercussions. For example, through gene therapy, a lab was able to make rats lose weight, but the long-term effects of the gene manipulation lead to worry of toxin production and too much weight loss. To prevent some of these issues from arising, scientists work towards stabilizing the entire process to make it safer before applying a higher degree of gene modification to the human embryos in IVF.

Autonomy and Tissue Ownership

For those who believe that human life begins at the moment of conception, this belief also suggests that human rights are given at that time. If human rights are given in this embryonic stage, then a surplus of ethical issues arise from manipulating the embryo in the realm of tissue ownership. In the long run, if implanted into a female and born, the embryo becomes an adult and has to now live with the genetic modifications chosen for them through the process of IVF. Unfortunately, in this base, cellular state, consent for gene manipulation is impossible. This leads to decision making by the parents. Rightful parental ownership over the embryo is only in the short-run and means that they control the embryos biological future. Consent over tissue ownership has been an issue for decades and can have legal repercussions. In the case of Henrietta Lacks, researchers lacked patient consent to use her tissues in genetic research, and this led to many legal issues on the family's right to profit from the use of her cells. Decisiveness over autonomy is necessary in the case of IVF to avoid long run issues and give people their full rights of humanity.

Profit Desire of the Industry

Many people do not oppose the IVF practice itself (i.e. the creating of a pregnancy through "artificial" ways) but are highly critical of the current state of the present day industry. Such individuals argue that the industry has now become a multibillion-dollar industry, which is widely unregulated and prone to serious abuses in the desire of practitioners to obtain profit. For instance, in 2008, a California physician transferred 12 embryos to a woman who gave birth to octuplets. This has made international news, and had led to accusations that many doctors are willing to seriously endanger the health and even life of women in order to gain money. Robert Winston, professor of fertility studies at Imperial College London, had called the industry "corrupt" and "greedy" saying that "One of the major problems facing us in healthcare is that IVF has become a massive commercial industry," and that "What has happened, of course, is that money is corrupting this whole technology", and accused authorities of failing to protect couples from exploitation "The regulatory authority has done a consistently bad job. It's not prevented the exploitation of women, it's not put out very good information to couples, it's not limited the number of unscientific treatments people have access to". The IVF industry can thus be seen as an example of what social scientists are describing as an increasing trend towards a market-driven construction of health, medicine and the human body.

As the science progresses, the industry is further driven by money in that researchers and innovators enter into the fight over patents and intellectual property rights. The Copyright Clause in the US Constitution protects innovator's rights to their respective work in attempts to promote scientific progress. Essentially, this lawful protection gives incentive to the innovators by providing them a temporary monopoly over their respective work. In the IVF industry, already incredibly expensive for patients, patents risk even higher prices for the patients to receive the procedure as they have to also cover the costs of protected works. For example, company 23andMe has patented a process used to calculate probability of gene inheritance. While this innovation could help many, the company retains sole right to administer it and thus does not have economic competition. Lack of economic competition leads to higher prices of products.

The industry has been accused of making unscientific claims, and distorting facts relating to infertility, in particular through widely exaggerated claims about how common infertility is in society, in an attempt to get as many couples as possible and as soon as possible to try treatments (rather than trying to conceive naturally for a longer time). This risks removing infertility from its social context and reducing the experience to a simple biological malfunction, which not only *can* be treated through bio-medical procedures, but *should* be treated by them. Indeed, there are serious concerns about the overuse of treatments, for instance Dr. Sami David, a fertility specialist and one of the pioneers of the early days of the IVF treatments, has expressed disappointment over the current state of the industry, and said many procedures are unnecessary; he said: "It's being the first choice of treatment rather than the last choice. When it was first opening up in late 1970s, early 80s, it was meant to be the last resort. Now it's a first resort. I think that's an injustice to women. I also think it can harm women in the long run." IVF thus raises ethical issues concerning the abuse of bio-medical facts to 'sell' corrective procedures and treatments for conditions that deviate from a constructed ideal of the 'healthy' or 'normal' body i.e., fertile females and males with reproductive systems capable of co-producing offspring.

Pregnancy Past Menopause

Although menopause is a natural barrier to further conception, IVF has allowed women to be pregnant in their fifties and sixties. Women whose uteruses have been appropriately prepared receive embryos that originated from an egg of an egg donor. Therefore, although these women do not have a genetic link with the child, they have an emotional link through pregnancy and childbirth. In many cases the genetic father of the child is the woman's partner. Even after menopause the uterus is fully capable of carrying out a pregnancy.

Allowing women to get pregnant past the natural time can factor into issues of overpopulation. Through the PGD, children born through IVF would credibly have higher life expectancy rates due to eliminated diseases. So increasing the amount of women who are able to bear children increases the population growth rate, while PGD in IVF decreases the death rate, resulting in an increasing population.

Same-sex Couples, Single and Unmarried Parents

A 2009 statement from the ASRM found no persuasive evidence that children are harmed or disadvantaged solely by being raised by single parents, unmarried parents, or homosexual parents. It did not support restricting access to assisted reproductive technologies on the basis of a prospective parent's marital status or sexual orientation.

Ethical concerns include reproductive rights, the welfare of offspring, nondiscrimination against unmarried individuals, homosexual, and professional autonomy.

A recent controversy in California focused on the question of whether physicians opposed to same-sex relationships should be required to perform IVF for a lesbian couple. Guadalupe T. Benitez, a lesbian medical assistant from San Diego, sued doctors Christine Brody and Douglas Fenton of the North Coast Women's Care Medical Group after Brody told her that she had "religious-based objections to treating her and homosexuals in general to help them conceive children by artificial insemination," and Fenton refused to authorise a refill of her prescription for the fertility drug Clomid on the same grounds. The California Medical Association had initially sided with Brody and Fenton, but the case, North Coast Women's Care Medical Group v. Superior Court, was decided unanimously by the California State Supreme Court in favor of Benitez on 19 August 2008.

Nadya Suleman came to international attention after having twelve embryos implanted, eight of which survived, resulting in eight newborns being added to her existing six-child family. The Medical Board of California sought to have fertility doctor Michael Kamrava, who treated Suleman, stripped of his license. State officials allege that performing Suleman's procedure is evidence of unreasonable judgment, substandard care, and a lack of concern for the eight children she would conceive and the six she was already struggling to raise. On 1 June 2011 the Medical Board issued a ruling that Kamrava's medical license be revoked effective 1 July 2011.

Anonymous Donors

Some children conceived by IVF using anonymous donors report being troubled over not knowing about their donor parent as well any genetic relatives they may have and their family history.

Alana Stewart, who was conceived using donor sperm, began an online forum for donor children called AnonymousUS in 2010. The forum welcomes the viewpoints of anyone involved in the IVF process. Olivia Pratten, a donor-conceived Canadian, sued the province of British Columbia for access to records on her donor father's identity in 2008. "I'm not a treatment, I'm a person, and those records belong to me," Pratten said. In May 2012, a court ruled in Pratten's favor, agreeing that the laws at the time discriminated against donor children and making anonymous sperm and egg donation in British Columbia illegal.

In the U.K., Sweden, Norway, Germany, Italy, New Zealand, and some Australian states, donors are not paid and cannot be anonymous.

In 2000, a website called Donor Sibling Registry was created to help biological children with a common donor connect with each other.

In 2012, a documentary called *Anonymous Father's Day* was released that focuses on donor-conceived children.

Unwanted Embryos

During the selection and transfer phases many embryos may be discarded in favour of others. This selection may be based on criteria such as genetic disorders or the sex. One of the earliest cases of special gene selection through IVF was the case of the Collins family in the 1990s, who selected the sex of their child. The ethic issues remain unresolved as no consensus exists in science, religion, and philosophy on when a human embryo should be recognized as a person. For those who believe that this is at the moment of conception, IVF becomes a moral question when multiple eggs are fertilised, begin development, and only a few are chosen for implantation.

If IVF were to involve the fertilisation of only a single egg, or at least only an amount that will be implanted, then this would not be an issue. However, this has the chance of increasing costs dramatically as only a few eggs can be attempted at a time. As a result, the couple must decide what to do with these extra embryos. Depending on their view of the embryo's humanity or the chance the couple will want to try to have another child, the couple has multiple options for dealing with these extra embryos. Couples can choose to keep them frozen, donate them to other infertile couples, thaw them, or donate them to medical research. Keeping them frozen costs money, donating them does not ensure they will survive, thawing them renders them immediately unviable, and medical research results in their termination. In the realm of medical research, the couple is not necessarily told what the embryos will be used for, and as a result, some can be used in stem cell research, a field perceived to have ethical issues.

Religious Response

The Roman Catholic Church opposes all kinds of assisted reproductive technology and artificial contraception, asserting they separate the procreative goal of marital sex from the goal of uniting married couples. The Roman Catholic Church permits the use of a small number of reproductive technologies and contraceptive methods like natural family planning, which involves charting ovulation times. The church allows other forms of reproductive technologies that allow conception to take place from normative sexual intercourse, such as a fertility lubricant. Pope Benedict XVI had publicly re-emphasised the Catholic Church's opposition to *in vitro* fertilisation, claiming it replaces love between a husband and wife. The Catechism of the Catholic Church claims that Natural law teaches that reproduction has an "inseparable connection" to sexual union of married couples. In addition, the church opposes IVF because it might cause disposal of embryos; in Catholicism, an embryo is viewed as an individual with a soul that must be treated as a person. The Catholic Church maintains that it is not objectively evil to be infertile, and advocates adoption as an option for such couples who still wish to have children.

Hindus welcomed the IVF as gift for those who can't bear child and termed doctors related to IVF doing punya as there are several characters who were claimed to be born without intercourse, mainly Karna and five Pandavas.

Regarding the response to IVF of Islam, the conclusions of Gad El-Hak Ali Gad El-Hak's ART fatwa include that:

- IVF of an egg from the wife with the sperm of her husband and the transfer of the fertilised egg back to the uterus of the wife is allowed, provided that the procedure is indicated for a medical reason and is carried out by an expert physician.

- Since marriage is a contract between the wife and husband during the span of their marriage, no third party should intrude into the marital functions of sex and procreation. This means that a third party donor is not acceptable, whether he or she is providing sperm, eggs, embryos, or a uterus. The use of a third party is tantamount to *zina*, or adultery.

Within the Orthodox Jewish community the concept is debated as there is little precedent in traditional Jewish legal textual sources. Regarding laws of sexuality, religious challenges include masturbation (which may be regarded as "seed wasting"), laws related to sexual activity and menstruation (niddah) and the specific laws regarding intercourse. An additional major issue is that of establishing paternity and lineage. For a baby conceived naturally, the father's identity is determined by a legal presumption (chazakah) of legitimacy: *rov bi'ot achar ha'baal* - a woman's sexual relations are assumed to be with her husband. Regarding an IVF child, this assumption does not exist and as such Rabbi Eliezer Waldenberg (among others) requires an outside supervisor to positively identify the father. Reform Judaism has generally approved *in vitro* fertilisation.

Society and Culture

Many people of sub-Saharan Africa choose to foster their children to infertile women. IVF enables these infertile women to have their own children, which imposes new ideals to a culture in which fostering children is seen as both natural and culturally important. Many infertile women are able to earn more respect in their society by taking care of the children of other mothers, and this may be lost if they choose to use IVF instead. As IVF is seen as unnatural, it may even hinder their societal position as opposed to making them equal with fertile women. It is also economically advantageous for infertile women to raise foster children as it gives these children greater ability to access resources that are important for their development and also aids the development of their society at large. If IVF becomes more popular without the birth rate decreasing, there could be more large family homes with fewer options to send their newborn children. This could result in an increase of orphaned children and/or a decrease in resources for the children of large families. This would ultimately stifle the children's and the community's growth.

Emotional Involvement

Studies have indicated that IVF mothers show greater emotional involvement with their child, and they enjoy motherhood more than mothers by natural conception. Similarly, studies have indicated that IVF fathers express more warmth and emotional involvement than fathers by adoption and natural conception and enjoy fatherhood more. Some IVF parents become overly involved with their children.

Men and IVF

Research has shown that men largely view themselves as 'passive' contributors since they have 'less physical involvement' in IVF treatment. Despite this, many men feel distressed after seeing the toll of hormonal injections and ongoing physical intervention on their partner. Fertility was found to be a significant factor in a man's perception of his masculinity, driving many to keep the treatment a secret. In cases where the men did share that he and his partner were undergoing IVF, they reported to have been teased, mainly by other men, although some viewed this as an affirmation of support and friendship. For others, this led to feeling socially isolated. In comparison with women, men showed less deterioration in mental health in the years following a failed treatment. However many men did feel guilt, disappointment and inadequacy, stating that they were simply trying to provide an 'emotional rock' for their partners.

Availability and Utilisation

High costs keep IVF out of reach for many developing countries, but research by the Genk Institute for Fertility Technology, in Belgium, claim to have found a much lower cost methodology (about 90% reduction) with similar efficacy, which may be suitable for some fertility treatment. Moreover, the laws of many countries permit IVF for only single women, lesbian couples, and persons participating in surrogacy arrangements. Using PGD gives members of these select demographic groups disproportionate access to a means of creating a child possessing characteristics that they consider "ideal," raising issues of equal opportunity for both the parents'/parent's and the child's generation. Many fertile couples now demand equal access to embryonic screening so that their child can be just as healthy as one created through IVF. Mass use of PGD, especially as a means of population control or in the presence of legal measures related to population or demographic control, can lead to intentional or unintentional demographic effects such as the skewed live-birth sex ratios seen in communist China following implementation of its one-child policy.

USA

In the USA, overall availability of IVF in 2005 was 2.5 IVF physicians per 100,000 population, and utilisation was 236 IVF cycles per 100,000. Utilisation highly increases with availability and IVF insurance coverage, and to a significant extent also with percentage of single persons and median income. In the USA 126 procedures are performed per million people per year. In the USA an average cycle, from egg retrieval to embryo implantation, costs $12,400, and insurance companies that do cover treatment, even partially, usually cap the number of cycles they pay for.

The cost of IVF rather reflects the costliness of the underlying healthcare system than the regulatory or funding environment, and ranges, on average for a standard IVF cycle and in 2006 United States dollars, between $12,500 in the United States to $4,000 in Japan. In Ireland, IVF costs around €4,000, with fertility drugs, if required, costing up to €3,000. The cost per live birth is highest in the United States ($41,000) and United Kingdom ($40,000) and lowest in Scandinavia and Japan (both around $24,500).

Many fertility clinics in the United States limit the upper age at which women are eligible for IVF to 50 or 55 years. These cut-offs make it difficult for women older than fifty-five to utilise the procedure.

Australia

In Australia, the average age of women undergoing ART treatment is 35.5 years among those using their own eggs (one in four being 40 or older) and 40.5 years among those using donated eggs.

Cameroon

Ernestine Gwet Bell supervised the first Cameroonian child born through IVF in 1998.

Israel

Israel has the highest rate of IVF in the world, with 1657 procedures performed per million people per year. The second highest rate is in Iceland, with 899 procedures per million people per year. Israel provides unlimited free in vitro procedures for its citizens for up to two children per woman under 45 years of age. In other countries the coverage of such procedures is limited if it exists at all. The Israeli Health Ministry says it spends roughly $3450 per procedure.

United Kingdom

Availability of IVF in England is determined by Clinical commissioning groups. The National Institute for Health and Care Excellence recommends up to 3 cycles of treatment for women under 40 and one cycle for some women aged between 40 and 42, but financial pressure has eroded compliance with this recommendation. CCGs in Essex, Bedfordshire and Somerset have reduced funding to one cycle, or none and it is expected that reductions will become more widespread. Funding may be available in "exceptional circumstances" – for example if a male partner has a transmittable infection or one partner is affected by cancer treatment. According to the campaign group Fertility Fairness at the end of 2014 every CCG in England was funding at least one cycle of IVF". Prices paid by the NHS in England varied between under £3,000 to more than £6,000 in 2014/5.

Legal Status

Government agencies in China passed bans on the use of IVF in 2003 by unmarried women or by couples with certain infectious diseases.

Sunni Muslim nations generally allow IVF between married couples when conducted with their own respective sperm and eggs, but not with donor eggs from other couples. But Iran, which is Shi'a Muslim, has a more complex scheme. Iran bans sperm donation but allows donation of both fertilised and unfertilised eggs. Fertilised eggs are donated from married couples to other married couples, while unfertilised eggs are donated in the context of mut'ah or temporary marriage to the father.

Costa Rica has a complete ban on IVF technology, it having been ruled unconstitutional by the nation's Supreme Court because it "violated life." Costa Rica is the only country in the western hemisphere that forbids IVF. A law project sent reluctantly by the government of President Laura Chinchilla was rejected by parliament. President Chinchilla has not publicly stated her position on the question of IVF. However, given the massive influence of the Catholic Church in her government any change in the status quo seems very unlikely. In spite of Costa Rican government and strong religious opposition, the IVF ban has been struck down by the Inter-American Court of Human Rights in a decision of 20 December 2012. The court said that a long-standing Costa Rican guarantee of protection for every human embryo violated the reproductive freedom of infertile couples because it prohibited them from using IVF, which often involves the disposal of embryos not implanted in a patient's uterus. On 10 September 2015, President Luis Guillermo Solís signed a decree legalising in-vitro fertilisation. The decree was added to the country's official gazette on 11 September. Opponents of the practice have since filed a lawsuit before the country's Constitutional Court.

All major restrictions on single but infertile women using IVF were lifted in Australia in 2002 after a final appeal to the Australian High Court was rejected on procedural grounds in the Leesa Meldrum case. A Victorian federal court had ruled in 2000 that the existing ban on all single women and lesbians using IVF constituted sex discrimination. Victoria's government announced changes to its IVF law in 2007 eliminating remaining restrictions on fertile single women and lesbians, leaving South Australia as the only state maintaining them.

Federal regulations in the United States include screening requirements and restrictions on donations, but generally do not affect sexually intimate partners. However, doctors may be required to *provide* treatments due to nondiscrimination laws, as for example in California. The US state of Tennessee proposed a bill in 2009 that would have defined donor IVF as adoption. During the same session another bill proposed barring adoption from any unmarried and cohabiting couple, and activist groups stated that passing the first bill would effectively stop unmarried people from using IVF. Neither of these bills passed.

Genome

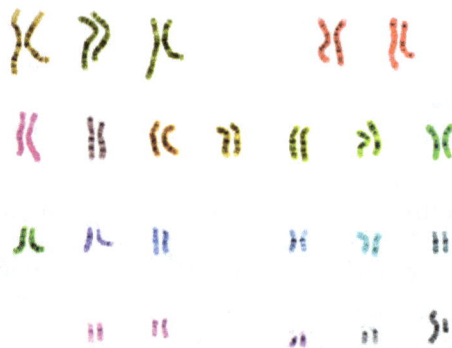

An image of the 46 chromosomes making up the diploid genome of a human male.
(The mitochondrial chromosome is not shown.)

In modern molecular biology and genetics, a genome is the genetic material of an organism. It consists of DNA (or RNA in RNA viruses). The genome includes both the genes, (the coding regions), the noncoding DNA and the genomes of the mitochondria and chloroplasts.

Origin of Term

The term was created in 1920 by Hans Winkler, professor of botany at the University of Hamburg, Germany. The Oxford Dictionary suggests the name to be a blend of the words *gene* and *chromosome*. A few related *-ome* words already existed—such as *biome, rhizome*, forming a vocabulary into which *genome* fits systematically.

Overview

Some organisms have multiple copies of chromosomes: diploid, triploid, tetraploid and so on. In classical genetics, in a sexually reproducing organism (typically eukarya) the gamete has half the number of chromosomes of the somatic cell and the genome is a full set of chromosomes in a diploid cell. The halving of the genetic material in gametes is accomplished by the segregation of homologous chromosomes during meiosis. In haploid organisms, including cells of bacteria, archaea, and in organelles including mitochondria and chloroplasts, or viruses, that similarly contain genes, the single or set of circular or linear chains of DNA (or RNA for some viruses), likewise constitute the genome. The term *genome* can be applied specifically to mean what is stored on a complete set of nuclear DNA (i.e., the «nuclear genome") but can also be applied to what is stored within organelles that contain their own DNA, as with the "mitochondrial genome" or the "chloroplast genome". Additionally, the genome can comprise non-chromosomal genetic elements such as viruses, plasmids, and transposable elements.

Typically, when it is said that the genome of a sexually reproducing species has been "sequenced", it refers to a determination of the sequences of one set of autosomes and one of each type of sex chromosome, which together represent both of the possible sexes. Even in species that exist in only one sex, what is described as a "genome sequence" may be a composite read from the chromosomes of various individuals. Colloquially, the phrase "genetic makeup" is sometimes used to signify the genome of a particular individual or organism. The study of the global properties of genomes of related organisms is usually referred to as genomics, which distinguishes it from genetics which generally studies the properties of single genes or groups of genes.

Both the number of base pairs and the number of genes vary widely from one species to another, and there is only a rough correlation between the two (an observation known as the C-value paradox). At present, the highest known number of genes is around 60,000, for the protozoan causing trichomoniasis, almost three times as many as in the human genome.

An analogy to the human genome stored on DNA is that of instructions stored in a book:

- The book (genome) would contain 23 chapters (chromosomes);

- Each chapter contains 48 to 250 million letters (A,C,G,T) without spaces;

- Hence, the book contains over 3.2 billion letters total;

- The book fits into a cell nucleus the size of a pinpoint;

- At least one copy of the book (all 23 chapters) is contained in most cells of our body. The only exception in humans is found in mature red blood cells which become enucleated during development and therefore lack a genome.

Sequencing and Mapping

```
CATGACGTCGCGGACAACCCAGAATTGTCTTGAGCGATGGTAAGATCTAACCTCACTGCCGGGGGAGGCTCATAC
CTGGGGCTTTACTGATGTCATACCGTCTTGCACGGGGATAGAATGACGGTGCCCGTGTCTGCTTGCCTCGAAGCA
ATTTTCTGAAAGTTACAGACTTCGATTAAAAAGATCGGACTGCGCGTGGGCCCGGAGAGACATGCGTGGTAGTCA
TTTTTCGACGTGTCAAGGACTCAAGGGAATAGTTTGGCGGGAGCGTTACAGCTTCAATTCCCAAAGGTCGCAAGA
CGATAAAATTCAACTACTGGTTTCGGCCTAATAGGTCACGTTTTATGTGAAATAGAGGGGAACCGGCTCCCAAAT
CCCTGGGTGTTCTATGATAAGTCCTGCTTTATAACACGGGGCGGTTAGGTTAAATGACTCTTCTATCTTATGGTG
ATCCAAGCGCCCGCTAATTCTGTTCTGTTAATGTTCATACCAATACTCACATCACATTAGATCAAAGGATCCCCG
AGCCCAGTCGCAAGGGTCTGCTGCTGTTGTCGACGCCTCATGTTACTCCTGGAATCTACCTGCCCTCCCCTCACC
GGTTAAGGCGTGTGATCGACGATGCAGGTATACATCGGCTCGGACCTACAGTGGTCGATCGACTGGCTACTGGCT
TCGCGGTTCGGCGCGTAGTTGAGTGCGATAACCCAACCGGTGGCAAGTAGCAAGAAGACCTACCTGGGTCACCTT
AGACAACCTAACTAATAGTCTCTAACGGGGAATTACCTTTACCAGTCTCATGCCTCCAATATATCTGCACCGCTT
CAATGATATCGCCCACAGAAAGTAGGGTCTCAGGTATCGCATACGCCGCGCCCGGGTCCCAGCTACGCTCAGGAC
GACAGTAGAGAGCTATTGTGTAATTCAGGCTCAGCATTCATCGACCTTTCCTGTTGTGAATATTGTGCTAATGCA
TCTCGTCCGTAACGATCTGGGGGGCAAAACCGAATATCCGTATTCTCGTCCTACGGGTCCACAATGAGAAAGTCC
TGCGCGTGATCGTCAGTTAAGTTAAATTAATTCAGGCTACGGTAAACTTGTAGTGAGCTAAGAATCACGGGAATC
ACGGGTTCGCTACAGATGAACTGAATTTATACACGGACAACTCATCGCCCATTTGGGCGTGGGCACCGCAGATCA
AAAGTGGCAGATTAGGAGTGCTTGATCAGGTTAGCAGGTGGACTGTATCCAACAGCGCATCAAACTTCAATAAAT
CCAAAGCGTTGTAGTGGTCTAAGCACCCCTGAACAGTGGCGCCCATCGTTAGCGTAGTACAACCCTTCCCCCTTG
AGGTGCGACATGGGGCCAGTTAGCCTGCCCTATATCCCTTGCACACGTTCAATAAGAGGGGCTCTACAGCGCCGC
TTTTTAAATTAGGATGCCGACCCCATCATTGGTAACTGTATGTTCATAGATATTTCTTCAGGAGTAATAGCGACA
AGCTGACACGCAAGGGTCAACAATAATTTCTACTATCACCCCGCTGAACGACTGTCTTTGCAAGAACCAACTGGG
CTTAGATTCGCGTCCTAACGTAGTGAGGGCCGAGTCATCATCAGATCAGGCATGAGAAACCGACGTCGAGTCTA
CACACGAGTTGTAAACAACTTGATTGCTATACTGTAGCTACCGCAAGGATCTCCTACATCAAAGACTACTGGGCG
ATCTGGATCCGAGTCAGAAATACGAGTTAATGCAAATTTACGTAGACCGGTGAAAACACGTGCCATGGGTTGCGT
AGACCGTAGTCAGAAGTGTGGCGCGCTATTCGTACCGAACCGGTGGAGTATACAGAATTGCTCTTCTACGACGTA
AGGAGCTCGGTCCCCAATGCACGCCAAAAAAGGAATAAAGTATTCAAACTGCGCATGGTCCCTCCGCCGGTGGCA
CTATTATCCATCCGAACGTTGAACCTACTTCCTCGGCTTATGCTGTCCTCAACAGTATCGCTTATGAATCGCATG
CGGCTGTGGATCTTAACGGCCACATTCTTAATTCCGACCGATCACCGATCGCCTTTCCTCGCTGGTACAATGAGT
ACTAAGTTATCCAGATCAAGGTTTGAACGGACTCGTATGACATGTGTGACTGAACCCGGGAGGAAATGCAGAGAA
CTGTTTCAAGGCCTCTGCTTTGGTATCACTCAATATATTCAGACCAGACAAGTGGCAAAATTTCGTGCGCCTCTC
CTAGGTATTCACGCAACCGTCGTAACATGCACTAAGGATAACTAGCGCCAGGGGGGCATACTAGGTCCCGGAGCT
AAAGACTACCCTATGGATTCCTTGGAGCGGGGACAATGCAGACCGGTTACGACACAATTATCGGGATCGTCTAGA
GGTATTATTAGCAAGACAATAAAGGACATTGCACAGAGACTTATTAGAATTCAACAAACAGGATCATATCATGCG
GTGTTGGGTCGGGCAAGTCCCCGAAGCTCGGCCAAAAGATTCGCCATGGAACCGTCTGGTCCTGTTAGCGTGTAC
GCCTGCTCCTGTTCCGGGTACCATAGATAGACTGAGATTGCGTCAAAAAATTGCGGCGAAAATAGAGGGGCTCCT
TGTAGAAATACCAGACTGGGGAATTTAAGCGCTTTCCACTATCTGAGCGACTAAACATCAACAAATGCGTCTACT
CGAATCCGCAGTAGGCAATTACAACCTGGTTCAGATCACTGGTTAATCAGGGATGTCTTCATAAGATTATACTTG
CCCCGACGCGACAGCTCTTCAAGGGGCCGATTTTTGGACTTCAGATACGCTAGAATTTAAAGGGTCTCTTACACC
TGCTGCGGCCTGCAGGGACCCCTAGAACTTGCCGCCTACTTGTCTCAGTCTAATAACGCGCGAAGCCGTGGGGCA
CGTGACCTTAAGTCGCAGAGCGAGTGATGAATTTGGGACGCTAATATGGGTGAATAGAGACTTATATCATCAGGG
```

Part of DNA sequence - prototypification of complete genome of virus

In 1976, Walter Fiers at the University of Ghent (Belgium) was the first to establish the complete nucleotide sequence of a viral RNA-genome (Bacteriophage MS2). The next year Fred Sanger completed the first DNA-genome sequence: Phage Φ-X174, of 5386 base pairs. The first complete genome sequences among all three domains of life were released within a short period during the mid-1990s: The first bacterial genome to be sequenced was that of Haemophilus influenzae, completed by a team at The Institute for Genomic Research in 1995. A few months later, the first eukaryotic genome was completed, with sequences of the 16 chromosomes of budding yeast *Saccharomyces cerevisiae* published as the result of a European-led effort begun in the mid-1980s. The first genome sequence for an archaeon, *Methanococcus jannaschii*, was completed in 1996, again by The Institute for Genomic Research.

The development of new technologies has made it dramatically easier and cheaper to do sequencing, and the number of complete genome sequences is growing rapidly. The US National Institutes of Health maintains one of several comprehensive databases of genomic information. Among the thousands of completed genome sequencing projects include those for rice, a mouse, the plant *Arabidopsis thaliana*, the puffer fish, and the bacteria E. coli. In December 2013, scientists first sequenced the entire *genome* of a Neanderthal, an extinct species of humans. The genome was extracted from the toe bone of a 130,000-year-old Neanderthal found in a Siberian cave.

New sequencing technologies, such as massive parallel sequencing have also opened up the prospect of personal genome sequencing as a diagnostic tool, as pioneered by Manteia Predictive Medicine. A major step toward that goal was the completion in 2007 of the full genome of James D. Watson, one of the co-discoverers of the structure of DNA.

Whereas a genome sequence lists the order of every DNA base in a genome, a genome map identifies the landmarks. A genome map is less detailed than a genome sequence and aids in navigating around the genome. The Human Genome Project was organized to map and to sequence the human genome. A fundamental step in the project was the release of a detailed genomic map by Jean Weissenbach and his team at the Genoscope in Paris.

Genome Compositions

Genome composition is used to describe the make up of contents of a haploid genome, which should include genome size, proportions of non-repetitive DNA and repetitive DNA in details. By comparing the genome compositions between genomes, scientists can better understand the evolutionary history of a given genome.

When talking about genome composition, one should distinguish between prokaryotes and eukaryotes as there are significant differences with contents structure. In prokaryotes, most of the genome (85–90%) is non-repetitive DNA, which means coding DNA mainly forms it, while non-coding regions only take a small part. On the contrary, eukaryotes have the feature of exon-intron organization of protein coding genes; the variation of repetitive DNA content in eukaryotes is also extremely high. In mammals and plants, the major part of the genome is composed of repetitive DNA.

Most biological entities that are more complex than a virus sometimes or always carry additional genetic material besides that which resides in their chromosomes. In some contexts, such as sequencing the genome of a pathogenic microbe, "genome" is meant to include information stored on this auxiliary material, which is carried in plasmids. In such circumstances then, "genome" describes all of the genes and information on non-coding DNA that have the potential to be present.

In eukaryotes such as plants, protozoa and animals, however, "genome" carries the typical connotation of only information on chromosomal DNA. So although these organisms contain chloroplasts or mitochondria that have their own DNA, the genetic information contained by DNA within these organelles is not considered part of the genome. In fact, mitochondria are sometimes said to have their own genome often referred to as the "mitochondrial genome". The DNA found within the chloroplast may be referred to as the "plastome".

Genome Size

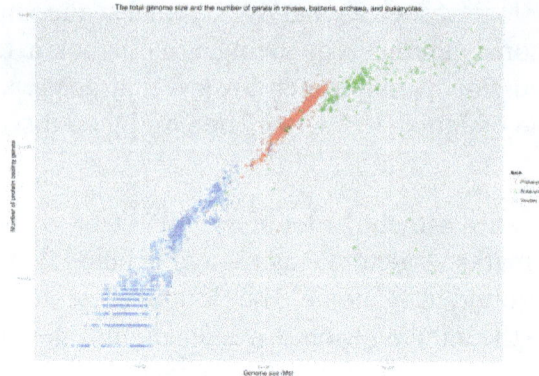

Log-log plot of the total number of annotated proteins in genomes submitted to GenBank as a function of genome size.

Genome size is the total number of DNA base pairs in one copy of a haploid genome. The genome size is positively correlated with the morphological complexity among prokaryotes and lower eukaryotes; however, after mollusks and all the other higher eukaryotes above, this correlation is no longer effective. This phenomenon also indicates the mighty influence coming from repetitive DNA act on the genomes.

Since genomes are very complex, one research strategy is to reduce the number of genes in a genome to the bare minimum and still have the organism in question survive. There is experimental work being done on minimal genomes for single cell organisms as well as minimal genomes for multi-cellular organisms. The work is both *in vivo* and *in silico*.

Here is a table of some significant or representative genomes.

Organism type	Organism	Genome size (base pairs)		Approx. no. of genes	Note
Virus	Porcine circovirus type 1	1,759	1.8kb		Smallest viruses replicating autonomously in eukaryotic cells.
Virus	Bacteriophage MS2	3,569	3.5kb		First sequenced RNA-genome
Virus	SV40	5,224	5.2kb		
Virus	Phage Φ-X174	5,386	5.4kb		First sequenced DNA-genome

Virus	HIV	9,749	9.7kb		
Virus	Phage λ	48,502	48kb		Often used as a vector for the cloning of re-combinant DNA.
Virus	Megavirus	1,259,197	1.3Mb		Until 2013 the largest known viral genome.
Virus	*Pandoravirus salinus*	2,470,000	2.47Mb		Largest known viral genome.
Bacterium	*Nasuia delto-cephalinicola* (strain NAS-ALF)	112,091	112kb		Smallest non-viral genome.
Bacterium	*Carsonella ruddii*	159,662	160kb		
Bacterium	*Buchnera aphid-icola*	600,000	600kb		
Bacterium	*Wigglesworthia glossinidia*	700,000	700Kb		
Bacterium	*Haemophilus influenzae*	1,830,000	1.8Mb		First genome of a living organism sequenced, July 1995
Bacterium	*Escherichia coli*	4,600,000	4.6Mb	4288	
Bacterium	*Solibacter usita-tus* (strain Ellin 6076)	9,970,000	10Mb		
Bacterium – cyanobacterium	*Prochlorococcus spp.* (1.7 Mb)	1,700,000	1.7Mb	1884	Smallest known cyano-bacterium genome
Bacterium – cyanobacterium	*Nostoc puncti-forme*	9,000,000	9Mb	7432	7432 "open reading frames"
Amoeboid	*Polychaos dubi-um ("Amoeba" dubia)*	670,000,000,000	670Gb		Largest known genome. (Disputed)
Plant	*Genlisea tubero-sa*	61,000,000	61Mb		Smallest recorded flow-ering plant genome, 2014.
Plant	*Arabidopsis thaliana*	157,000,000	157Mb	25498	First plant genome sequenced, December 2000.
Plant	*Populus tricho-carpa*	480,000,000	480Mb	73013	First tree genome sequenced, September 2006
Plant	*Fritillaria as-syrica*	130,000,000,000	130Gb		

Plant	*Paris japonica* (Japanese-native, pale-petal)	150,000,000,000	150Gb		Largest plant genome known
Plant – moss	*Physcomitrella patens*	480,000,000	480Mb		First genome of a bryophyte sequenced, January 2008.
Fungus – yeast	*Saccharomyces cerevisiae*	12,100,000	12.1Mb	6294	First eukaryotic genome sequenced, 1996
Fungus	*Aspergillus nidulans*	30,000,000	30Mb	9541	
Nematode	*Pratylenchus coffeae*	20,000,000	20Mb		Smallest animal genome known
Nematode	*Caenorhabditis elegans*	100,300,000	100Mb	19000	First multicellular animal genome sequenced, December 1998
Insect	*Drosophila melanogaster* (fruit fly)	175,000,000	175Mb	13600	Size variation based on strain (175-180Mb; standard $y\ w$ strain is 175Mb)
Insect	*Apis mellifera* (honey bee)	236,000,000	236Mb	10157	
Insect	*Bombyx mori* (silk moth)	432,000,000	432Mb	14623	14,623 predicted genes
Insect	*Solenopsis invicta* (fire ant)	480,000,000	480Mb	16569	
Mammal	*Mus musculus*	2,700,000,000	2.7Gb	20210	
Mammal	*Homo sapiens*	3,200,000,000	3.2Gb	20000	*Homo sapiens* estimated genome size 3.2 billion bp Initial sequencing and analysis of the human genome
Mammal	*Bonobo*	3,286,640,000	3.3Gb	20000	*Pan paniscus* estimated genome size 3.29 billion bp
Fish	*Tetraodon nigroviridis* (type of puffer fish)	385,000,000	390Mb		Smallest vertebrate genome known estimated to be 340 Mb – 385 Mb.
Fish	*Protopterus aethiopicus* (marbled lungfish)	130,000,000,000	130Gb		Largest vertebrate genome known

Proportion of Non-repetitive DNA

The proportion of non-repetitive DNA is calculated by using the length of non-repetitive DNA divided by genome size. Protein-coding genes and RNA-coding genes are generally non-repetitive DNA. A bigger genome does not mean more genes, and the proportion of non-repetitive DNA decreases along with increasing genome size in higher eukaryotes.

It had been found that the proportion of non-repetitive DNA can vary a lot between species. Some *E. coli* as prokaryotes only have non-repetitive DNA, lower eukaryotes such as *C. elegans* and fruit fly, still possess more non-repetitive DNA than repetitive DNA. Higher eukaryotes tend to have more repetitive DNA than non-repetitive ones. In some plants and amphibians, the proportion of non-repetitive DNA is no more than 20%, becoming a minority component.

Proportion of Repetitive DNA

The proportion of repetitive DNA is calculated by using length of repetitive DNA divide by genome size. There are two categories of repetitive DNA in genome: tandem repeats and interspersed repeats.

Tandem Repeats

Tandem repeats are usually caused by slippage during replication, unequal crossing-over and gene conversion, satellite DNA and microsatellites are forms of tandem repeats in the genome. Although tandem repeats count for a significant proportion in genome, the largest proportion in mammalian is the other type, interspersed repeats.

Interspersed Repeats

Interspersed repeats mainly come from transposable elements (TEs), but they also include some protein coding gene families and pseudogenes. Transposable elements are able to integrate into the genome at another site within the cell. It is believed that TEs are an important driving force on genome evolution of higher eukaryotes. TEs can be classified into two categories, Class 1 (retrotransposons) and Class 2 (DNA transposons).

Retrotransposons

Retrotransposons can be transcribed into RNA, which are then duplicated at another site into the genome. Retrotransposons can be divided into Long terminal repeats (LTRs) and Non-Long Terminal Repeats (Non-LTR).

Long Terminal Repeats (LTRs)

> similar to retroviruses, which have both gag and pol genes to make cDNA from RNA and proteins to insert into genome, but LTRs can only act within the cell

as they lack the env gene in retroviruses. It has been reported that LTRs consist of the largest fraction in most plant genome and might account for the huge variation in genome size.

Non-Long Terminal Repeats (Non-LTRs)

can be divided into long interspersed elements (LINEs), short interspersed elements (SINEs) and Penelope-like elements. In *Dictyostelium discoideum*, there is another DIRS-like elements belong to Non-LTRs. Non-LTRs are widely spread in eukaryotic genomes.

Long interspersed elements (LINEs)

are able to encode two Open Reading Frames (ORFs) to generate transcriptase and endonuclease, which are essential in retrotransposition. The human genome has around 500,000 LINEs, taking around 17% of the genome.

Short interspersed elements (SINEs)

are usually less than 500 base pairs and need to co-opt with the LINEs machinery to function as nonautonomous retrotransposons. The Alu element is the most common SINEs found in primates, it has a length of about 350 base pairs and takes about 11% of the human genome with around 1,500,000 copies.

DNA Transposons

DNA transposons generally move by "cut and paste" in the genome, but duplication has also been observed. Class 2 TEs do not use RNA as intermediate and are popular in bacteria, in metazoan it has also been found.

Genome Evolution

Genomes are more than the sum of an organism's genes and have traits that may be measured and studied without reference to the details of any particular genes and their products. Researchers compare traits such as *chromosome number* (karyotype), genome size, gene order, codon usage bias, and GC-content to determine what mechanisms could have produced the great variety of genomes that exist today.

Duplications play a major role in shaping the genome. Duplication may range from extension of short tandem repeats, to duplication of a cluster of genes, and all the way to duplication of entire chromosomes or even entire genomes. Such duplications are probably fundamental to the creation of genetic novelty.

Horizontal gene transfer is invoked to explain how there is often extreme similarity between small portions of the genomes of two organisms that are otherwise very distantly related. Horizontal gene transfer seems to be common among many microbes. Also, eukaryotic cells seem to have experienced a transfer of some genetic material from their chloroplast and mitochondrial genomes to their nuclear chromosomes.

Genome Project

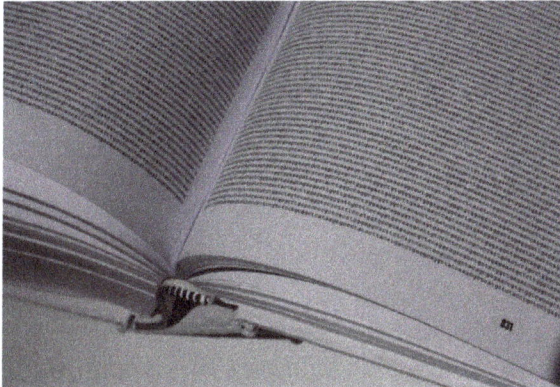

When printed, the human genome sequence fills around 100 huge books of close print

Genome projects are scientific endeavours that ultimately aim to determine the complete genome sequence of an organism (be it an animal, a plant, a fungus, a bacterium, an archaean, a protist or a virus) and to annotate protein-coding genes and other important genome-encoded features. The genome sequence of an organism includes the collective DNA sequences of each chromosome in the organism. For a bacterium containing a single chromosome, a genome project will aim to map the sequence of that chromosome. For the human species, whose genome includes 22 pairs of autosomes and 2 sex chromosomes, a complete genome sequence will involve 46 separate chromosome sequences.

The Human Genome Project was a landmark genome project that is already having a major impact on research across the life sciences, with potential for spurring numerous medical and commercial developments.

Genome Assembly

Genome assembly refers to the process of taking a large number of short DNA sequences and putting them back together to create a representation of the original chromosomes from which the DNA originated. In a shotgun sequencing project, all the DNA from a source (usually a single organism, anything from a bacterium to a mammal) is first fractured into millions of small pieces. These pieces are then "read" by automated sequencing machines, which can read up to 1000 nucleotides or bases at a time.

(The four bases are adenine, guanine, cytosine, and thymine, represented as AGCT.) A genome assembly algorithm works by taking all the pieces and aligning them to one another, and detecting all places where two of the short sequences, or *reads*, overlap. These overlapping reads can be merged, and the process continues.

Genome assembly is a very difficult computational problem, made more difficult because many genomes contain large numbers of identical sequences, known as repeats. These repeats can be thousands of nucleotides long, and some occur in thousands of different locations, especially in the large genomes of plants and animals.

The resulting (draft) genome sequence is produced by combining the information sequenced contigs and then employing linking information to create scaffolds. Scaffolds are positioned along the physical map of the chromosomes creating a "golden path".

Assembly Software

Originally, most large-scale DNA sequencing centers developed their own software for assembling the sequences that they produced. However, this has changed as the software has grown more complex and as the number of sequencing centers has increased. An example of such assembler *Short Oligonucleotide Analysis Package* developed by BGI for de novo assembly of human-sized genomes, alignment, SNP detection, resequencing, indel finding, and structural variation analysis.

Genome Annotation

Genome annotation is the process of attaching biological information to sequences. It consists of three main steps:

1. identifying portions of the genome that do not code for proteins

2. identifying elements on the genome, a process called gene prediction, and

3. attaching biological information to these elements.

Automatic annotation tools try to perform all this by computer analysis, as opposed to manual annotation (a.k.a. curation) which involves human expertise. Ideally, these approaches co-exist and complement each other in the same annotation pipeline.

The basic level of annotation is using BLAST for finding similarities, and then annotating genomes based on that. However, nowadays more and more additional information is added to the annotation platform. The additional information allows manual annotators to deconvolute discrepancies between genes that are given the same annotation. Some databases use genome context information, similarity scores, experimental data, and integrations of other resources to provide genome annotations through their Subsystems approach. Other databases (e.g. Ensembl) rely on both curated data sources as well as a range of different software tools in their automated genome annotation pipeline.

Structural annotation consists of the identification of genomic elements.

- ORFs and their localisation

- gene structure

- coding regions

- location of regulatory motifs

Functional annotation consists of attaching biological information to genomic elements.

- biochemical function

- biological function

- involved regulation and interactions

- expression

These steps may involve both biological experiments and *in silico* analysis. Proteogenomics based approaches utilize information from expressed proteins, often derived from mass spectrometry, to improve genomics annotations.

A variety of software tools have been developed to permit scientists to view and share genome annotations.

Genome annotation remains a major challenge for scientists investigating the human genome, now that the genome sequences of more than a thousand human individuals and several model organisms are largely complete. Identifying the locations of genes and other genetic control elements is often described as defining the biological "parts list" for the assembly and normal operation of an organism. Scientists are still at an early stage in the process of delineating this parts list and in understanding how all the parts "fit together".

Genome annotation is an active area of investigation and involves a number of different organizations in the life science community which publish the results of their efforts in publicly available biological databases accessible via the web and other electronic means. Here is an alphabetical listing of on-going projects relevant to genome annotation:

- Encyclopedia of DNA elements (ENCODE)

- Entrez Gene

- Ensembl

- GENCODE

- Gene Ontology Consortium

- GeneRIF

- RefSeq

- Uniprot

- Vertebrate and Genome Annotation Project (Vega)

When is a Genome Project Finished?

When sequencing a genome, there are usually regions that are difficult to sequence (often regions with highly repetitive DNA). Thus, 'completed' genome sequences are rarely ever complete, and terms such as 'working draft' or 'essentially complete' have been used to more accurately describe the status of such genome projects. Even when every base pair of a genome sequence has been determined, there are still likely to be errors present because DNA sequencing is not a completely accurate process. It could also be argued that a complete genome project should include the sequences of mitochondria and (for plants) chloroplasts as these organelles have their own genomes.

It is often reported that the goal of sequencing a genome is to obtain information about the complete set of genes in that particular genome sequence. The proportion of a genome that encodes for genes may be very small (particularly in eukaryotes such as humans, where coding DNA may only account for a few percent of the entire sequence). However, it is not always possible (or desirable) to only sequence the coding regions separately. Also, as scientists understand more about the role of this noncoding DNA (often referred to as junk DNA), it will become more important to have a complete genome sequence as a background to understanding the genetics and biology of any given organism.

In many ways genome projects do not confine themselves to only determining a DNA sequence of an organism. Such projects may also include gene prediction to find out where the genes are in a genome, and what those genes do. There may also be related projects to sequence ESTs or mRNAs to help find out where the genes actually are.

Historical and Technological Perspectives

Historically, when sequencing eukaryotic genomes (such as the worm *Caenorhabditis elegans*) it was common to first map the genome to provide a series of landmarks across the genome. Rather than sequence a chromosome in one go, it would be sequenced piece by piece (with the prior knowledge of approximately where that piece is located on the larger chromosome). Changes in technology and in particular improvements to the processing power of computers, means that genomes can now be 'shotgun sequenced' in one go (there are caveats to this approach though when compared to the traditional approach).

Improvements in DNA sequencing technology has meant that the cost of sequencing a new genome sequence has steadily fallen (in terms of cost per base pair) and newer technology has also meant that genomes can be sequenced far more quickly.

When research agencies decide what new genomes to sequence, the emphasis has been on species which are either high importance as model organism or have a relevance to human health (e.g. pathogenic bacteria or vectors of disease such as mosquitos) or species which have commercial importance (e.g. livestock and crop plants). Secondary emphasis is placed on species whose genomes will help answer important questions in molecular evolution (e.g. the common chimpanzee).

In the future, it is likely that it will become even cheaper and quicker to sequence a genome. This will allow for complete genome sequences to be determined from many different individuals of the same species. For humans, this will allow us to better understand aspects of human genetic diversity.

Example Genome Projects

L1 Dominette 01449, the Hereford who serves as the subject of the Bovine Genome Project

Many organisms have genome projects that have either been completed or will be completed shortly, including:

- Humans, Homo sapiens;

- Humans, Homo sapiens;

- Palaeo-Eskimo, an ancient-human

- Neanderthal, "*Homo neanderthalensis*" (partial);

- Common chimpanzee *Pan troglodytes*;

- Domestic Cow

- Bovine Genome

- Honey Bee Genome Sequencing Consortium

- Horse genome

- Human microbiome project

- International Grape Genome Program

- International HapMap Project

- Tomato 150+ genome resequencing project

- 100K Genome Project

- Genomics England

Metabolome

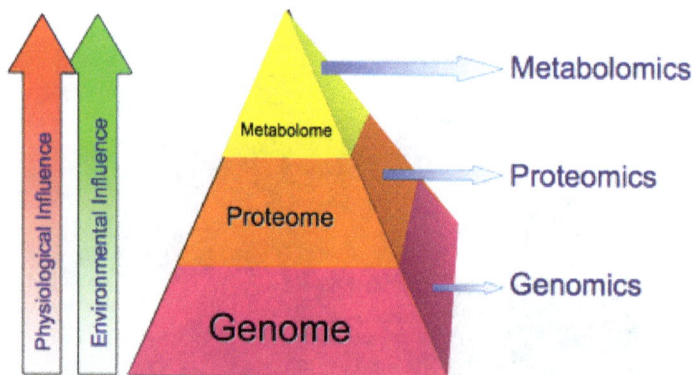

Figure. An illustration of the "pyramid of life" which shows the varying influence of the environment and physiology on the genome, proteome and metabolome.

The metabolome refers to the complete set of small-molecule chemicals found within a biological sample. The biological sample can be a cell, a cellular organelle, an organ, a tissue, a tissue extract, a biofluid or an entire organism. The small molecule chemicals found in a given metabolome may include both endogenous metabolites that are naturally produced by an organism (such as amino acids, organic acids, nucleic acids, fatty acids, amines, sugars, vitamins, co-factors, pigments, antibiotics, etc.) as well as exogenous chemicals (such as drugs, environmental contaminants, food additives, toxins and other xenobiotics) that are not naturally produced by an organism. In other words, there is both an endogenous metabolome and an exogenous metabolome. The endogenous metabolome can be further subdivided to include a "primary" and a "secondary" metabolome (particularly when referring to plant or microbial metabolomes). A primary metabolite is directly involved in the normal growth, development, and reproduction. A secondary metabolite is not directly involved in those processes, but usually has important ecological function. Secondary metabolites may include pigments, antibiotics or waste products derived from partially metabolized xenobiotics. To qualify as a metabolite, or to be considered to be part of the metabolome, a small molecule must typically have a molecular weight <1500 Da. This means that molecules such as glycolipids, polysaccharides, short peptides (<14 amino acids) and small oligonucleotides (<5 bases) can be regarded as metabolites or constituents of the metabolome. On the other hand, very large macromolecules such as proteins, messenger RNA, ribosomal RNA, microRNA and DNA are definitely not metabolites and are not considered to be part of the metabolome. The study of the metabolome is called metabolomics. Figure for a picture of the relationship between different "omes".

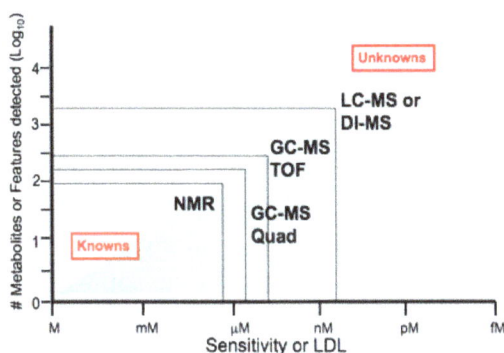

Figure . An illustration of the relationship between sensitivity (or lower detection limit in concentration units) and the number of metabolites detected (using a Log10 scale) via different metabolomics technologies.

Origins

The word metabolome appears to be a blending of the words "metabolite" and "chromosome". It was constructed to imply that metabolites are indirectly encoded by genes or act on genes and gene products. The term "metabolome" was first used in 1998 and was likely coined to match with existing biological terms referring to the complete set of genes (the genome), the complete set of proteins (the proteome) and the complete set of transcripts (the transcriptome). The first book on metabolomics was published in 2003. The first journal dedicated to metabolomics (titled simply "Metabolomics") was launched in 2005 and is currently edited by Dr. Royston Goodacre. Some of the more significant early papers on metabolome analysis are listed in the references below

Measuring the Metabolome

The metabolome reflects the interaction between an organism's genome and its environment. As a result, an organism's metabolome can serve as an excellent probe of its phenotype (i.e. the product of its genotype and its environment). Metabolites can be measured (identified, quantified or classified) using a number of different technologies including NMR spectroscopy and mass spectrometry. Most mass spectrometry (MS) methods must be coupled to various forms of liquid chromatography (LC), gas chromatography (GC) or capillary electrophoresis (CE) to facilitate compound separation. Each method is typically able to identify or characterize 50-5000 different metabolites or metabolite "features" at a time, depending on the instrument or protocol being used. Currently it is not possible to analyze the entire range of metabolites by a single analytical method.

Nuclear magnetic resonance (NMR) spectroscopy is an analytical chemistry technique that measures the absorption of radiofrequency radiation of specific nuclei when molecules containing those nuclei are placed in strong magnetic fields. The frequency (i.e. the chemical shift) at which a given atom or nucleus absorbs is highly dependent on the chemical environment (bonding, chemical structure nearest neighbours, solvent) of that atom in a given molecule. The NMR absorption patterns produce "resonance" peaks at different frequencies or different chemical shifts – this collection of peaks is called a NMR spectrum. Because each chemical compound has a different chemical structure, each compound will have a unique (or almost unique) NMR spectrum. As a result, NMR is particularly useful for the characterization, identification and quantification of small molecules, such as metabolites. The widespread use of NMR for "classical" metabolic studies, along with its exceptional capacity to handle complex metabolite mixtures is likely the reason why NMR was one of the first technologies to be widely adopted for routine metabolome measurements. As an analytical technique, NMR is non-destructive, non-biased, easily quantifiable, requires little or no separation, permits the identification of novel compounds and it needs no chemical derivatization. NMR is particularly amenable to detecting compounds that are less tractable to LC-MS analysis, such as sugars, amines or volatile liquids or GC-MS analysis, such as large molecules (>500 Da) or relatively non-reactive compounds. NMR is not a very sensitive technique with a lower limit of detection of about 5 μM. Typically 50-150 compounds can be identified by NMR-based metabolomic studies.

Mass spectrometry is an analytical technique that measures the mass-to-charge ratio of molecules. Molecules or molecular fragments are typically charged or ionized by spraying them through a charged field (electrospray ionization), bombarding them with electrons from a hot filament (electron ionization) or blasting them with a laser when they are placed on specially coated plates (matrix assisted laser desorption ionization). The charged molecules are then propelled through space using electrodes or magnets and their speed, rate of curvature, or other physical characteristics are measured to determine their mass-to-charge ratio. From these data the mass of the parent molecule can be determined. Further fragmentation of the molecule through controlled collisions with gas molecules or with electrons can help determine the structure of molecules. Very accurate mass measurements can also be used to determine the elemental formulas or elemental composition of compounds. Most forms of mass spectrometry require some form of separation using liquid chromatography or gas chromatography. This separation step is required to simplify the resulting mass spectra and to permit more accurate compound identification. Some mass spectrometry methods also require that the molecules be derivatized or chemically modified so that they are more amenable for chromatographic separation (this is particularly true for GC-MS). As an analytical technique, MS is a very sensitive method that requires very little sample (<1 ng of material or <10 μL of a biofluid) and can generate signals for 1000s of metabolites from a single sample. MS instruments can also be configured for very high throughput metabolome analyses (100s to 1000s of samples a day). Quantification of metabolites and the characterization of novel compound structures is more difficult by MS than by NMR. LC-MS is particularly amenable to detecting hydrophobic molecules (lipids, fatty acids) and peptides while GC-MS is best for detecting small molecules (<500 Da) and highly volatile compounds (esters, amines, ketones, alkanes, thiols).

Unlike the genome or even the proteome, the metabolome is a highly dynamic entity that can change dramatically, over a period of just seconds or minutes. As a result, there is growing interest in measuring metabolites over multiple time periods or over short time intervals using modified versions of NMR or MS-based metabolomics

Metabolome Databases

Because an organism's metabolome is largely defined by its genome, different species will have different metabolomes. Indeed, the fact that the metabolome of a tomato is different than the metabolome of an apple is the reason why these two fruits taste so different. Furthermore, different tissues, different organs and biofluids associated with those organs and tissues can also have distinctly different metabolomes. The fact that different organisms and different tissues/biofluids have such different metabolomes has led to the development of a number of organism-specific of biofluid-specific metabolome databases. Some of the better known metabolome databases include the Human Metabolome Database or HMDB, the Yeast Metabolome Database or YMDB, the E. coli Metabolome Database or ECMDB, the Arabidopsis metabolome database or AraCyc as well as the Urine Metabolome Database, the Cerebrospinal Fluid (CSF) Metabolome Database and the Serum Metabolome Database. The latter three databases are specific to human biofluids. A number of very popular general metabolite databases also exist including KEGG, MetaboLights, the Golm Metabolome Database, MetaCyc, LipidMaps and Metlin. Metabolome databases can be distinguished from metabolite databases in that metabolite databases contain lightly annotated or synoptic metabolite data from multiple organisms while metabolome databases contain richly detailed and heavily referenced chemical, pathway, spectral and metabolite concentration data for specific organisms.

The Human Metabolome Database

The Human Metabolome Database is a freely available, open-access database containing detailed data on more than 40,000 metabolites that have already been identified or are likely to be found in the human body. The HMDB contains three kinds of information: 1) chemical information, 2) clinical information, and 3) biochemical information. The chemical data includes >40,000 metabolite structures with detailed descriptions, extensive chemical classifications, synthesis information and observed/calculated chemical properties. It also contains nearly 10,000 experimentally measured NMR, GC-MS and LC/MS spectra from more than 1100 different metabolites. The clinical information includes data on >10,000 metabolite-biofluid concentrations, metabolite concentration information on more than 600 different human diseases and pathway data for more than 200 different inborn errors of metabolism. The biochemical information includes nearly 6000 protein (and DNA) sequences and more than 5000 biochemical reactions that are linked to these metabolite entries. The HMDB supports a wide variety of online queries including text searches, chemical structure searches, sequence similarity searches and spectral similarity searches. This makes it particularly useful for metabolomic researchers who are attempting to identify or understand metabolites in clinical metabolomic studies. The first version of the HMDB was released in Jan. 1 2007 and was compiled by scientists at the University of Alberta and the University of Calgary. At that time they reported data on 2,500 metabolites, 1,200 drugs and 3,500 food components. Since then these scientists have greatly expanded the collection. The latest version of the HMDB (version 3.5) contains >16,000 endogenous metabolites, >1500 drugs and >22,000 food constituents or food metabolites.

Human Biofluid Metabolomes

Scientists at the University of Alberta have been systematically characterizing specific biofluid metabolomes including the serum metabolome, the urine metabolome, the cerebrospinal fluid (CSF) metabolome and the saliva metabolome. These efforts have involved both experimental metabolomic analysis (involving NMR, GC-MS, ICP-MS, LC-MS and HPLC assays) as well as extensive literature mining. According to their data, the human serum metabolome contains at least 4200 different compounds (including many lipids), the human urine metabolome contains at least 3000 different compounds (including 100s of volatiles and gut microbial metabolites), the human CSF metabolome contains nearly 500 different compounds while the human saliva metabolome contains approximately 400 different metabolites, including many bacterial products.

Yeast Metabolome Database

The Yeast Metabolome Database is a freely accessible, online database of >2000 small molecule metabolites found in or produced by Saccharomyces cerevisiae (Baker's yeast). The YMDB contains two kinds of information: 1) chemical information and 2) biochemical information. The chemical information in YMDB includes 2027 metabolite structures with detailed metabolite descriptions, extensive chemical classifications, synthesis information and observed/calculated chemical properties. It also contains nearly 4000 NMR, GC-MS and LC/MS spectra obtained from more than 500 different metabolites. The biochemical information in YMDB includes >1100 protein (and DNA) sequences and >900 biochemical reactions. The YMDB supports a wide variety of queries including text searches, chemical structure searches, sequence similarity searches and spectral similarity searches. This makes it particularly useful for metabolomic researchers who are studying yeast as a model organism or who are looking into optimizing the production of fermented beverages (wine, beer).

The Escherichia Coli Metabolome Database

The E. Coli Metabolome Database is a freely accessible, online database of >2700 small molecule metabolites found in or produced by Escherichia coli (E. coli strain K12, MG1655). The ECMDB contains two kinds of information: 1) chemical information and 2) biochemical information. The chemical information includes more than 2700 metabolite structures with detailed metabolite descriptions, extensive chemical classifications, synthesis information and observed/calculated chemical properties. It also contains nearly 5000 NMR, GC-MS and LC-MS spectra from more than 600 different metabolites. The biochemical information includes >1600 protein (and DNA) sequences and >3100 biochemical reactions that are linked to these metabolite entries. The ECMDB supports many different types of online queries including text searches, chemical structure searches, sequence similarity searches and spectral similarity searches. This makes it particularly useful for metabolomic researchers who are studying E. coli as a model organism.

Human Metabolome Database

The Human Metabolome Database (HMDB) is a comprehensive, high-quality, freely accessible, online database of small molecule metabolites found in the human body. Created by the Human Metabolome Project funded by Genome Canada. One of the first dedicated metabolomics databases, the HMDB facilitates human metabolomics research, including the identification and characterization of human metabolites using NMR spectroscopy, GC-MS spectrometry and LC/MS spectrometry. To aid in this discovery process, the HMDB contains three kinds of data: 1) chemical data, 2) clinical data, and 3) molecular biology/biochemistry data (Fig. 1-3). The chemical data includes 41,514 metabolite structures with detailed descriptions along with nearly 10,000 NMR, GC-MS and LC/MS spectra.

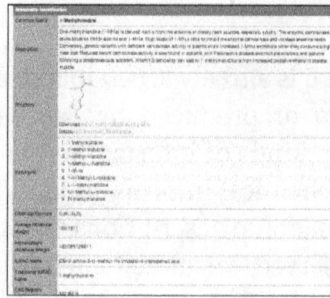

Fig. 1. HMDB metabolite

Fig. 2. HMDB metabolite chemical and physical information

Fig. 3. HMDB metabolite clinical information

The clinical data includes information on >10,000 metabolite-biofluid concentrations and metabolite concentration information on more than 600 different human diseases. The biochemical data includes 5,688 protein (and DNA) sequences and more than 5000 biochemical reactions that are linked to these metabolite entries. Each metabolite entry in the HMDB contains more than 110 data fields with 2/3 of the information being devoted to chemical/clinical data and the other 1/3 devoted to enzymatic or biochemical data. Many data fields are hyperlinked to other databases (KEGG, MetaCyc, PubChem, Protein Data Bank, ChEBI, Swiss-Prot, and GenBank) and a variety of structure and pathway viewing applets. The HMDB database supports extensive text, sequence, spectral, chemical structure and relational query searches. It has been widely used in metabolomics, clinical chemistry, biomarker discovery and general biochemistry education.

Four additional databases, DrugBank, T3DB, SMPDB and FooDB are also part of the HMDB suite of databases. DrugBank contains equivalent information on ~1600 drug and drug metabolites, T3DB contains information on 3100 common toxins and environmental pollutants, SMPDB contains pathway diagrams for 700 human metabolic and disease pathways, while FooDB contains equivalent information on ~28,000 food components and food additives.

Version History

The first version of HMDB was released on January 1, 2007, followed by two subsequent versions on January 1, 2009 (version 2.0), August 1, 2009 (version 2.5), September 18, 2012 (version 3.0) and Jan. 1, 2013 (version 3.5). Details for each of the major HMDB versions (up to version 3.0) is provided in Table 1.

Table 1. Content comparison of HMDB 1.0 with HMDB 2.0 and HMDB 3.0			
Database Feature or Content Status	**HMDB (v1.0)**	**HMDB (v2.0)**	**HMDB (v3.0)**
Number of metabolites	2,180	6,408	37,170
Number of unique metabolite synonyms	27,700	43,882	152,364
Number of compounds with disease links	862	1,002	3,948
Number of compounds with biofluid or tissue concentration data	883	4,413	6,796
Number of compounds with chemical synthesis references	220	1,647	8,863
Number of compounds with experimental reference 1H and or 13C NMR spectra	385	792	1,054
Number of compounds with reference MS/MS spectra	390	799	1,249
Number of compounds with reference GC-MS reference data	0	279	884
Number of human-specific pathway maps	26	58	442
Number of compounds in Human Metabolome Library	607	920	1,031
Number of HMDB data fields	91	102	114
'Number of predicted molecular properties	2	2	10

Scope and Access

All data in HMDB is non-proprietary or is derived from a non-proprietary source. It is freely accessible and available to anyone. In addition, nearly every data item is fully traceable and explicitly referenced to the original source. HMDB data is available through a public web interface and downloads.

References

- Nice.org Fertility: Assessment and Treatment for People with Fertility Problems. London: RCOG Press. 2004. ISBN 1-900364-97-2.

- Schulman, Joseph D. (2010) Robert G. Edwards – A Personal Viewpoint, CreateSpace Independent Publishing Platform, ISBN 1456320750.

- Professor Henry Louis Gates, Jr.; Professor Emmanuel Akyeampong; Mr. Steven J. Niven (2 February 2012). Dictionary of African Biography. OUP USA. pp. 25–. ISBN 978-0-19-538207-5.

- Griffiths JF; Gelbart WM; Lewontin RC; Wessler SR; Suzuki DT; Miller JH (2005). Introduction to Genetic Analysis. New York: W.H. Freeman and Co. pp. 34–40, 473–476, 626–629. ISBN 0-7167-4939-4.

- Madigan M; Martinko J, eds. (2006). Brock Biology of Microorganisms (11th ed.). Prentice Hall. ISBN 0-13-144329-1.

- Ridley, M. (2006), Genome: the autobiography of a species in 23 chapters (PDF), New York: Harper Perennial, ISBN 0-06-019497-9.

- Mankertz P (2008). "Molecular Biology of Porcine Circoviruses". Animal Viruses: Molecular Biology. Caister Academic Press. ISBN 978-1-904455-22-6.

- Koonin, Eugene V. (2011-08-31). The Logic of Chance: The Nature and Origin of Biological Evolution. FT Press. ISBN 9780132542494.

- Stojanovic, edited by Nikola (2007). Computational genomics : current methods. Wymondham: Horizon Bioscience. ISBN 1-904933-30-0.

- Pevsner, Jonathan (2009). Bioinformatics and functional genomics (2nd ed.). Hoboken, N.J: Wiley-Blackwell. ISBN 9780470085851.

- Harrigan, G. G. & Goodacre, R. (eds) (2003). Metabolic Profiling: Its Role in Biomarker Discovery and Gene Function Analysis. Boston: Kluwer Academic Publishers. ISBN 1-4020-7370-4.

- Thomas-MacLean R et al. No Cookie-Cutter Response: Conceptualizing Primary Health Care. Retrieved 26 August 2014.

- Breuer, Howard (22 October 2010). "Octomom's Doctor Tearfully Apologizes, Admits Mistake". People. Retrieved 22 May 2012.

- "Fiscal Note, HB 2159 – SB 2136, from Tennessee General Assembly Fiscal Review Committee" (PDF). Retrieved 22 May 2012.

- "China Bars In-Vitro Fertilization for Pregnancy". Redorbit.com. 12 November 2003. Archived from the original on 2011-07-15. Retrieved 22 May 2012.

4

Applications of Biomedicine

Medical diagnosis, differential diagnosis, therapy and the 2009 flu pandemic vaccine are the significant topics included in this chapter. The process of determining a person's illness is known as medical diagnosis, while differential diagnosis is differentiating one disease from another that presents similar clinical features. The topics discussed in the chapter are of great importance to broaden the existing knowledge on biomedicine.

Medical Diagnosis

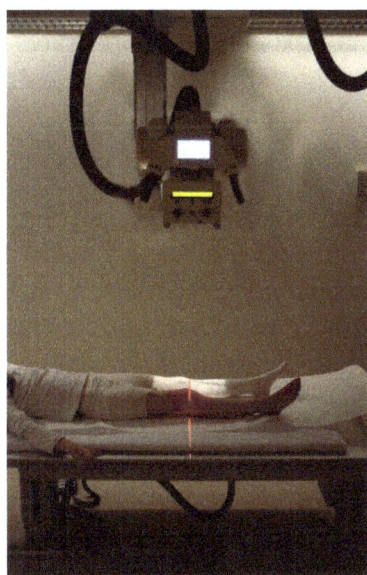

Radiography is an important tool in diagnosis of certain disorders.

Medical diagnosis (abbreviated Dx or D_S) is the process of determining which disease or condition explains a person's symptoms and signs. It is most often referred to as diagnosis with the medical context being implicit. The information required for diagnosis is typically collected from a history and physical examination of the person seeking medical care. Often, one or more diagnostic procedures, such as diagnostic tests, are also done during the process. Sometimes Posthumous diagnosis is considered a kind of medical diagnosis.

Diagnosis is often challenging, because many signs and symptoms are nonspecific. For example, redness of the skin (erythema), by itself, is a sign of many disorders and thus doesn't tell the healthcare professional what is wrong. Thus differential diagnosis, in which several possible explanations are compared and contrasted, must be performed. This involves the correlation of various pieces of information followed by the recognition and differentiation of patterns. Occasionally the process is made easy by a sign or symptom (or a group of several) that is pathognomonic.

Diagnosis is a major component of the procedure of a doctor's visit. From the point of view of statistics, the diagnostic procedure involves classification tests.

History

The first recorded examples of medical diagnosis are found in the writings of Imhotep (2630-2611 BC) in ancient Egypt (the Edwin Smith Papyrus). A Babylonian medical textbook, the *Diagnostic Handbook* written by Esagil-kin-apli (fl.1069-1046 BC), introduced the use of empiricism, logic and rationality in the diagnosis of an illness or disease. Traditional Chinese Medicine, as described in the Yellow Emperor's Inner Canon or Huangdi Neijing, specified four diagnostic methods: inspection, auscultation-olfaction, interrogation, and palpation. Hippocrates was known to make diagnoses by tasting his patients' urine and smelling their sweat.

Medical Uses

A diagnosis, in the sense of diagnostic procedure, can be regarded as an attempt at classification of an individual's condition into separate and distinct categories that allow medical decisions about treatment and prognosis to be made. Subsequently, a diagnostic opinion is often described in terms of a disease or other condition, but in the case of a wrong diagnosis, the individual's actual disease or condition is not the same as the individual's diagnosis.

A diagnostic procedure may be performed by various health care professionals such as a physician, physical therapist, optometrist, healthcare scientist, chiropractor, dentist, podiatrist, nurse practitioner, or physician assistant. This article uses *diagnostician* as any of these person categories.

A diagnostic procedure (as well as the opinion reached thereby) does not necessarily involve elucidation of the etiology of the diseases or conditions of interest, that is, what *caused* the disease or condition. Such elucidation can be useful to optimize treatment, further specify the prognosis or prevent recurrence of the disease or condition in the future.

The initial task is to detect a medical indication to perform a diagnostic procedure. Indications include:

- Detection of any deviation from what is known to be normal, such as can be described in terms of, for example, anatomy (the structure of the human body), physiology (how the body works), pathology (what can go wrong with the anatomy and physiology), psychology (thought and behavior) and human homeostasis (regarding mechanisms to keep body systems in balance). Knowledge of what is normal and measuring of the patient's current condition against those norms can assist in determining the patient's particular departure from homeostasis and the degree of departure, which in turn can assist in quantifying the indication for further diagnostic processing.

- A complaint expressed by a patient.

- The fact that a patient has sought a diagnostician can itself be an indication to perform a diagnostic procedure. For example, in a doctor's visit, the physician may already start performing a diagnostic procedure by watching the gait of the patient from the waiting room to the doctor's office even before she or he has started to present any complaints.

Even during an already ongoing diagnostic procedure, there can be an indication to perform another, separate, diagnostic procedure for another, potentially concomitant, disease or condition. This may occur as a result of an incidental finding of a sign unrelated to the parameter of interest, such as can occur in comprehensive tests such as radiological studies like magnetic resonance imaging or blood test panels that also include blood tests that are not relevant for the ongoing diagnosis.

Procedure

General components which are present in a diagnostic procedure in most of the various available methods include:

- Complementing the already given information with further data gathering, which may include questions of the medical history (potentially from other people close to the patient as well), physical examination and various diagnostic tests. A diagnostic test is any kind of medical test performed to aid in the diagnosis or detection of disease. Diagnostic tests can also be used to provide prognostic information on people with established disease.

- Processing of the answers, findings or other results. Consultations with other providers and specialists in the field may be sought.

There are a number of methods or techniques that can be used in a diagnostic procedure, including performing a differential diagnosis or following medical algorithms. In reality, a diagnostic procedure may involve components of multiple methods.

Differential Diagnosis

The method of differential diagnosis is based on finding as many candidate diseases or conditions as possible that can possibly cause the signs or symptoms, followed by a process of elimination or at least of rendering the entries more or less probable by further medical tests and other processing until, aiming to reach the point where only one candidate disease or condition remains as probable. The final result may also remain a list of possible conditions, ranked in order of probability or severity.

The resultant diagnostic opinion by this method can be regarded more or less as a diagnosis of exclusion. Even if it doesn't result in a single probable disease or condition, it can at least rule out any imminently life-threatening conditions.

Unless the provider is certain of the condition present, further medical tests, such as medical imaging, are performed or scheduled in part to confirm or disprove the diagnosis but also to document the patient's status and keep the patient's medical history up to date.

If unexpected findings are made during this process, the initial hypothesis may be ruled out and the provider must then consider other hypotheses.

Pattern Recognition

In a pattern recognition method the provider uses experience to recognize a pattern of clinical characteristics. It is mainly based on certain symptoms or signs being associated with certain diseases or conditions, not necessarily involving the more cognitive processing involved in a differential diagnosis.

This may be the primary method used in cases where diseases are "obvious", or the provider's experience may enable him or her to recognize the condition quickly. Theoretically, a certain pattern of signs or symptoms can be directly associated with a certain therapy, even without a definite decision regarding what is the actual disease, but such a compromise carries a substantial risk of missing a diagnosis which actually has a different therapy so it may be limited to cases where no diagnosis can be made.

Diagnostic criteria

The term *diagnostic criteria* designates the specific combination of signs, symptoms, and test results that the clinician uses to attempt to determine the correct diagnosis.

Some examples of diagnostic criteria, also known as clinical case definitions, are:

- Amsterdam criteria for hereditary nonpolyposis colorectal cancer

- McDonald criteria for multiple sclerosis

- ACR criteria for systemic lupus erythematosus

- Centor criteria for strep throat

Clinical Decision Support System

Clinical decision support systems are interactive computer programs designed to assist health professionals with decision-making tasks. The clinician interacts with the software utilizing both the clinician's knowledge and the software to make a better analysis of the patients data than either human or software could make on their own. Typically the system makes suggestions for the clinician to look through and the clinician picks useful information and removes erroneous suggestions.

Other Diagnostic Procedure Methods

Other methods that can be used in performing a diagnostic procedure include:

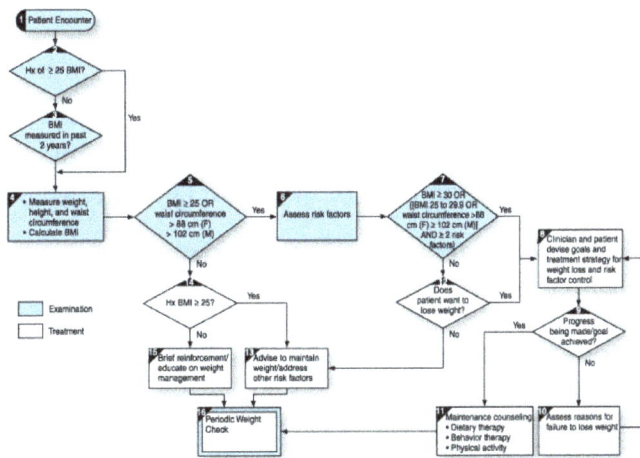

An example of a medical algorithm for assessment and treatment of overweight and obesity.

- Usage of medical algorithms

- An "exhaustive method", in which every possible question is asked and all possible data is collected.

Adverse Effects

Overdiagnosis

Overdiagnosis is the diagnosis of "disease" that will never cause symptoms or death during a patient's lifetime. It is a problem because it turns people into patients unnecessarily and because it can lead to economic waste (overutilization) and treatments that may cause harm. Overdiagnosis occurs when a disease is diagnosed correctly, but the diagnosis is irrelevant. A correct diagnosis may be irrelevant because treatment for the disease is not available, not needed, or not wanted.

Errors

Most people will experience at least one diagnostic error in their lifetime, according to a 2015 report by the National Academies of Sciences, Engineering, and Medicine.

Causes and factors of error in diagnosis are:

- the manifestation of disease are not sufficiently noticeable

- a disease is omitted from consideration

- too much significance is given to some aspect of the diagnosis

- the condition is a rare disease with symptoms suggestive of many other conditions

- the condition has a rare presentation

Lag Time

When making a medical diagnosis, a lag time is a delay in time until a step towards diagnosis of a disease or condition is made. Types of lag times are mainly:

- *Onset-to-medical encounter lag time*, the time from onset of symptoms until visiting a health care provider

- *Encounter-to-diagnosis lag time*, the time from first medical encounter to diagnosis

Society and Culture

Etymology

Medical diagnosis or the actual process of making a diagnosis is a cognitive process. A clinician uses several sources of data and puts the pieces of the puzzle together to make a diagnostic impression. The initial diagnostic impression can be a broad term describing a category of diseases instead of a specific disease or condition. After the initial diagnostic impression, the clinician obtains follow up tests and procedures to get more data to support or reject the original diagnosis and will attempt to narrow it down to a more specific level. Diagnostic procedures are the specific tools that the clinicians use to narrow the diagnostic possibilities.

Social Context

Diagnosis can take many forms. It might be a matter of naming the disease, lesion, dysfunction or disability. It might be a management-naming or prognosis-naming exercise. It may indicate either degree of abnormality on a continuum or kind of abnormality in a classification. It's influenced by non-medical factors such as power, ethics and financial incentives for patient or doctor. It can be a brief summation or an extensive formulation, even taking the form of a story or metaphor. It might be a means of communication such as a computer code through which it triggers payment, prescription, notification, information or advice. It might be pathogenic or salutogenic. It's generally uncertain and provisional.

Once a diagnostic opinion has been reached, the provider is able to propose a management plan, which will include treatment as well as plans for follow-up. From this point on, in addition to treating the patient's condition, the provider can educate the patient about the etiology, progression, prognosis, other outcomes, and possible treatments of her or his ailments, as well as providing advice for maintaining health.

A treatment plan is proposed which may include therapy and follow-up consultations and tests to monitor the condition and the progress of the treatment, if needed, usually according to the medical guidelines provided by the medical field on the treatment of the particular illness.

Relevant information should be added to the medical record of the patient.

A failure to respond to treatments that would normally work may indicate a need for review of the diagnosis.

Concepts Related to Diagnosis

Sub-types of diagnoses include:

Clinical diagnosis

> A diagnosis made on the basis of medical signs and patient-reported symptoms, rather than diagnostic tests

Laboratory diagnosis

> A diagnosis based significantly on laboratory reports or test results, rather than the physical examination of the patient. For instance, a proper diagnosis of infectious diseases usually requires both an examination of signs and symptoms, as well as laboratory characteristics of the pathogen involved.

Radiology diagnosis

> A diagnosis based primarily on the results from medical imaging studies. Greenstick fractures are common radiological diagnoses.

Principal diagnosis

> The single medical diagnosis that is most relevant to the patient's chief com-

plaint or need for treatment. Many patients have additional diagnoses.

Admitting diagnosis

The diagnosis given as the reason why the patient was admitted to the hospital; it may differ from the actual problem or from the *discharge diagnoses*, which are the diagnoses recorded when the patient is discharged from the hospital.

Differential diagnosis

A process of identifying all of the possible diagnoses that could be connected to the signs, symptoms, and lab findings, and then ruling out diagnoses until a final determination can be made.

Diagnostic criteria

Designates the combination of signs, symptoms, and test results that the clinician uses to attempt to determine the correct diagnosis. They are standards, normally published by international committees, and they are designed to offer the best sensitivity and specificity possible, respect the presence of a condition, with the state-of-the-art technology.

Prenatal diagnosis

Diagnosis work done before birth

Diagnosis of exclusion

A medical condition whose presence cannot be established with complete confidence from history, examination or testing. Diagnosis is therefore by elimination of all other reasonable possibilities.

Dual diagnosis

The diagnosis of two related, but separate, medical conditions or co-morbidities; the term almost always refers to a diagnosis of a serious mental illness and a substance addiction.

Self-diagnosis

The diagnosis or identification of a medical conditions in oneself. Self-diagnosis is very common and typically accurate for everyday conditions, such as headaches, menstrual cramps, and headlice.

Remote diagnosis

A type of telemedicine that diagnoses a patient without being physically in the same room as the patient.

Nursing diagnosis

> Rather than focusing on biological processes, a nursing diagnosis identifies people's responses to situations in their lives, such as a readiness to change or a willingness to accept assistance.

Computer-aided diagnosis

> Providing symptoms allows the computer to identify the problem and diagnose the user to the best of its ability. Health screening begins by identifying the part of the body where the symptoms are located; the computer cross-references a database for the corresponding disease and presents a diagnosis.

Overdiagnosis

> The diagnosis of "disease" that will never cause symptoms, distress, or death during a patient's lifetime

Wastebasket diagnosis

> A vague, or even completely fake, medical or psychiatric label given to the patient or to the medical records department for essentially non-medical reasons, such as to reassure the patient by providing an official-sounding label, to make the provider look effective, or to obtain approval for treatment. This term is also used as a derogatory label for disputed, poorly described, overused, or questionably classified diagnoses, such as pouchitis and senility, or to dismiss diagnoses that amount to overmedicalization, such as the labeling of normal responses to physical hunger as reactive hypoglycemia.

Retrospective diagnosis

> The labeling of an illness in a historical figure or specific historical event using modern knowledge, methods and disease classifications.

Differential Diagnosis

In medicine, a differential diagnosis is the distinguishing of a particular disease or condition from others that present similar clinical features. Differential diagnostic procedures are used by physicians and other trained medical professionals to diagnose the specific disease in a patient, or, at least, to eliminate any imminently life-threatening conditions. Often each individual option of a possible disease is called a differential diagnosis (for example, bronchitis could be a differential diagnosis in the evaluation of a cough that ends up with a final diagnosis of common cold).

More generally, a differential diagnostic procedure is a systematic diagnostic method used to identify the presence of a disease entity where multiple alternatives are possible. This method is essentially a process of elimination or at least a process of obtaining information that shrinks the "probabilities" of candidate conditions to negligible levels, by using evidence such as symptoms, patient history, and medical knowledge to adjust epistemic confidences in the mind of the diagnostician (or, for computerized or computer-assisted diagnosis, the software of the system).

Differential diagnosis can be regarded as implementing aspects of the hypothetico-deductive method, in the sense that the potential presence of candidate diseases or conditions can be viewed as hypotheses that physicians further determine as being true or false.

Common abbreviations of the term "differential diagnosis" include DDx, ddx, DD, D/Dx, or ΔΔ.

General Components

There are various methods of performing a differential diagnostic procedure, but in general, it is based on the idea that one begins by considering the most common diagnosis first: a head cold versus meningitis, for example. As a reminder, medical students are taught the Occam's razor adage, "When you hear hoofbeats, look for horses, not zebras," which means look for the simplest, most common explanation first. Only after ruling out the simplest diagnosis should the clinician consider more complex or exotic diagnoses.

Differential diagnosis has four steps. The physician:

1. Gathers all information about the patient and creates a symptoms list. The list can be in writing or in the physician's head, as long as they make a list.

2. Lists all possible causes (*candidate conditions*) for the symptoms. Again, this can be in writing or in the physician's head but it must be done.

3. Prioritizes the list by placing the most urgently dangerous possible causes at the top of the list.

4. Rules out or treats possible causes, beginning with the most urgently dangerous condition and working down the list. *Rule out*—practically—means use tests and other scientific methods to determine that a candidate condition has a clinically negligible probability of being the cause.

In some cases, there remains *no* diagnosis. This suggests the physician has made an error, or that the true diagnosis is unknown to medicine. The physician removes diagnoses from the list by observing and applying tests that produce different results, depending on which diagnosis is correct.

A mnemonic to help in considering multiple possible pathological processes is *VINDI-CATE'M*:

- Vascular

- Inflammatory/Infectious

- Neoplastic

- Degenerative/Defficiency/Drugs

- Idiopathic/Intoxication/Iatrogenic

- Congenital

- Autoimmune/Allergic/Anatomic

- Traumatic

- Endocrine/Environmental

- Metabolic

Specific Methods

There are several methods for differential diagnostic procedures, and several variants among those. Furthermore, a differential diagnostic procedure can be used concomitantly or alternately with protocols, guidelines, or other diagnostic procedures (such as pattern-recognition or using medical algorithms).

For example, in case of medical emergency, there may not be enough time to do any detailed calculations or estimations of different probabilities, in which case the ABC protocol (*Airway, Breathing and Circulation*) may be more appropriate. Later, when the situation is less acute, a more comprehensive differential diagnostic procedure may be adopted.

The differential diagnostic procedure may be simplified if a "pathognomonic" sign or symptom is found (in which case it is almost certain that the target condition is present) or in the absence of a *sine qua non* sign or symptom (in which case it is almost certain that the target condition is absent).

A diagnostician can be selective, considering first those disorders that are more likely (a probabilistic approach), more serious if left undiagnosed and untreated (a prognostic approach), or more responsive to treatment if offered (a pragmatic approach). Since the subjective probability of the presence of a condition is never exactly 100% or 0%, the differential diagnostic procedure may aim at specifying these various probabilities to form indications for further action.

The following are two methods of differential diagnosis, being based on epidemiology and likelihood ratios, respectively.

Epidemiology-based Method

One method of performing a differential diagnosis by epidemiology aims to estimate the probability of each candidate condition by comparing their probabilities to have occurred in the first place in the individual. It is based on probabilities related both to the presentation (such as pain) and probabilities of the various candidate conditions (such as diseases).

Theory

The statistical basis for differential diagnosis is Bayes' theorem. As an analogy, when a die has landed the outcome is certain by 100%, but the probability that it Would Have Occurred In the First Place (hereafter abbreviated WHOIFP) is still 1/6. In the same way, the probability that a presentation or condition would have occurred in the first place in an individual (WHOIFPI) is not same as the probability that the presentation or condition *has* occurred in the individual, because the presentation *has* occurred by 100% certainty in the individual. Yet, the contributive probability fractions of each condition are assumed the same, relatively:

$$\frac{\text{Pr(Presentation is caused by condition in individual)}}{\text{Pr(Presentation has occurred in individual)}} = \frac{\text{Pr(Presentation WHOIFPI by condition)}}{\text{Pr(Presentation WHOIFPI)}}$$

where:

- Pr(Presentation is caused by condition in individual) is the probability that the presentation is caused by condition in the individual

- *condition* without further specification refers to any candidate condition

- Pr(Presentation has occurred in individual) is the probability that the presentation has occurred in the individual, which can be perceived and thereby set at 100%

- Pr(Presentation WHOIFPI by condition) is the probability that the presentation Would Have Occurred in the First Place in the Individual by condition

- Pr(Presentation WHOIFPI) is the probability that the presentation Would Have Occurred in the First Place in the Individual

When an individual presents with a symptom or sign, Pr(Presentation has occurred in individual) is 100% and can therefore be replaced by 1, and can be ignored since division by 1 does not make any difference:

$$\text{Pr(Presentation is caused by condition in individual)} = \frac{\text{Pr(Presentation WHOIFPI by condition)}}{\text{Pr(Presentation WHOIFPI)}}$$

The total probability of the presentation to have occurred in the individual can be approximated as the sum of the individual candidate conditions:

$$
\begin{aligned}
\text{Pr(Presentation WHOIFPI)} \quad &= \text{Pr(Presentation WHOIFPI by condition 1)} \\
&+ \text{Pr(Presentation WHOIFPI by condition 2)} \\
&+ \text{Pr(Presentation WHOIFPI by condition 3)} + \text{etc.}
\end{aligned}
$$

Also, the probability of the presentation to have been caused by any candidate condition is proportional to the probability of the condition, depending on what rate it causes the presentation:

$$
\text{Pr(Presentation WHOIFPI by condition)} = \text{Pr(Condition WHOIFPI)} \cdot r_{\text{condition} \rightarrow \text{presentation}},
$$

where:

- Pr(Presentation WHOIFPI by condition) is the probability that the presentation Would Have Occurred in the First Place in the Individual by condition

- Pr(Condition WHOIFPI) is the probability that the condition Would Have Occurred in the First Place in the Individual

- $r_{\text{condition} \rightarrow \text{presentation}}$ is the rate for which a condition causes the presentation, that is, he fraction of people with condition that manifest with the presentation.

The probability that a condition would have occurred in the first place in an individual is approximately equal to that of a population that is as similar to the individual as possible except for the current presentation, compensated where possible by relative risks given by known risk factor that distinguish the individual from the population:

$$
\text{Pr(Condition WHOIFPI)} \approx RR_{\text{condition}} \cdot \text{Pr(Condition in population)},
$$

where:

- Pr(Condition WHOIFPI) is the probability that the condition Would Have Occurred in the First Place in the Individual

- $RR_{\text{condition}}$ is the relative risk for condition conferred by known risk factors in the individual that are not present in the population

- Pr(Condition in population) is the probability that the condition occurs in a population that is as similar to the individual as possible except for the presentation

The following table demonstrates how these relations can be made for a series of candidate conditions:

	Candidate condition 1	Candidate condition 2	Candidate condition 3
Pr(Condition in population)	Pr(Condition 1 in population)	Pr(Condition 2 in population)	Pr(Condition 3 in population)
$RR_{condition}$	RR_1	RR_2	RR_3
Pr(Condition WHOIFPI)	Pr(Condition 1 WHOIFPI)	Pr(Condition 2 WHOIFPI)	*P(Condition 3 WHOIFPI)*
$r_{Condition \rightarrow presentation}$	$r_{Condition\ 1 \rightarrow presentation}$	$r_{Condition\ 2 \rightarrow presentation}$	$r_{Condition\ 3 \rightarrow presentation}$
Pr(Presentation WHOIFPI by condition)	Pr(Presentation WHOIFPI by condition 1)	Pr(Presentation WHOIFPI by condition 2)	Pr(Presentation WHOIFPI by condition 3)
Pr(Presentation WHOIFPI) = the sum of the probabilities in row just above			
Pr(Presentation is caused by condition in individual)	Pr(Presentation is caused by condition 1 in individual)	Pr(Presentation is caused by condition 2 in individual)	Pr(Presentation is caused by condition 3 in individual)

One additional "candidate condition" is the instance of there being no abnormality, and the presentation is only a (usually relatively unlikely) appearance of a basically normal state. Its probability in the population (*P(No abnormality in population)*) is complementary to the sum of probabilities of "abnormal" candidate conditions.

Example

This example case demonstrates how this method is applied, but does not represent a guideline for handling similar real-world cases. Also, the example uses relatively specified numbers with sometimes several decimals, while in reality, there are often simply rough estimations, such as of likelihoods being *very high, high, low* or *very low*, but still using the general principles of the method.

For an individual (who becomes the "patient" in this example), a blood test of, for example, serum calcium shows a result above the standard reference range, which, by most definitions, classifies as hypercalcemia, which becomes the "presentation" in this case. A physician (who becomes the "diagnostician" in this example), who does not currently see the patient, gets to know about his finding.

By practical reasons, the physician considers that there is enough test indication to have a look at the patient's medical records. For simplicity, let's say that the only information given in the medical records is a family history of primary hyperparathyroidism (here abbreviated as PH), which may explain the finding of hypercalcemia. For this patient, let's say that the resultant hereditary risk factor is estimated to confer a relative risk of 10 ($RR_{PH} = 10$).

The physician considers that there is enough motivation to perform a differential diagnostic procedure for the finding of hypercalcemia. The main causes of hypercalcemia are primary hyperparathyroidism (PH) and cancer, so for simplicity, the list of candidate conditions that the physician could think of can be given as:

- Primary hyperparathyroidism (PH)

- Cancer

- Other diseases that the physician could think of (which is simply termed "other conditions" for the rest of this example)

- No disease (or no abnormality), and the finding is caused entirely by statistical variability

The probability that 'primary hyperparathyroidism' (PH) would have occurred in the first place in the individual ($P(PH\ WHOIFPI)$) can be calculated as follows:

Let's say that the last blood test taken by the patient was half a year ago and was normal, and that the incidence of primary hyperparathyroidism in a general population that appropriately matches the individual (except for the presentation and mentioned heredity) is 1 in 4000 per year. Ignoring more detailed retrospective analyses (such as including speed of disease progress and lag time of medical diagnosis), the time-at-risk for having developed primary hyperparathyroidism can roughly be regarded as being the last half-year, because a previously developed hypercalcemia would probably have been caught up by the previous blood test. This corresponds to a probability of primary hyperparathyroidism (PH) in the population of:

$$\text{Pr(PH in population)} = 0.5 \text{ years} \cdot \frac{1}{4000 \text{ per year}} = \frac{1}{8000}$$

With the relative risk conferred from the family history, the probability that primary hyperparathyroidism (PH) would have occurred in the first place in the individual given from the currently available information becomes:

$$\text{Pr(PH WHOIFPI)} \approx RR_{PH} \cdot \text{Pr(PH in population)} = 10 \cdot \frac{1}{8000} = \frac{1}{800} = 0.00125$$

Primary hyperparathyroidism can be assumed to cause hypercalcemia essentially 100% of the time ($r_{PH \rightarrow hypercalcemia} = 1$), so this independently calculated probability of primary hyperparathyroidism (PH) can be assumed to be the same as the probability of being a cause of the presentation:

$$\begin{aligned}\text{Pr(Hypercalcemia WHOIFPI by PH)} &= \text{Pr(PH WHOIFPI)} \cdot r_{PH \rightarrow hypercalcemia} \\ &= 0.00125 \cdot 1 = 0.00125\end{aligned}$$

For cancer, the same time-at-risk is assumed for simplicity, and let's say that the incidence of cancer in the area is estimated at 1 in 250 per year, giving a population probability of cancer of:

$$Pr(\text{cancer in population}) = 0.5 \text{ years} \cdot \frac{1}{250 \text{ per year}} = \frac{1}{500}$$

For simplicity, let's say that any association between a family history of primary hyperparathyroidism and risk of cancer is ignored, so the relative risk for the individual to have contracted cancer in the first place is similar to that of the population ($RR_{cancer} = 1$):

$$Pr(\text{cancer WHOIFPI}) \approx RR_{cancer} \cdot Pr(\text{cancer in population}) = 1 \cdot \frac{1}{500} = \frac{1}{500} = 0.002.$$

However, hypercalcemia only occurs in, very approximately, 10% of cancers, ($r_{cancer \to hypercalcemia} = 0.1$), so:

$$Pr(\text{Hypercalcemia WHOIFPI by cancer})$$
$$= Pr(\text{cancer WHOIFPI}) \cdot r_{cancer \to hypercalcemia}$$
$$= 0.002 \cdot 0.1 = 0.0002.$$

The probabilities that hypercalcemia would have occurred in the first place by other candidate conditions can be calculated in a similar manner. However, for simplicity, let's say that the probability that any of these would have occurred in the first place is calculated at 0.0005 in this example.

For the instance of there being no disease, the corresponding probability in the population is complementary to the sum of probabilities for other conditions:

$$Pr(\text{no disease in population}) = 1 - Pr(\text{PH in population}) - Pr(\text{cancer in population})$$
$$- Pr(\text{other conditions in population})$$
$$= 0.997.$$

The probability that the individual would be healthy in the first place can be assumed to be the same:

$$Pr(\text{no disease WHOIFPI}) = 0.997.$$

The rate at which the case of no abnormal condition still ends up in a measurement of serum calcium of being above the standard reference range (thereby classifying as hypercalcemia) is, by the definition of standard reference range, less than 2.5%. However, this probability can be further specified by considering how much the measurement deviates from the mean in the standard reference range. Let's say that the serum calcium measurement was 1.30 mmol/L, which, with a standard reference range established at

1.05 to 1.25 mmol/L, corresponds to a standard score of 3 and a corresponding probability of 0.14% that such degree of hypercalcemia would have occurred in the first place in the case of no abnormality:

$$r_{\text{no disease} \rightarrow \text{hypercalcemia}} = 0.0014$$

Subsequently, the probability that hypercalemia would have resulted from no disease can be calculated as:

$$\text{Pr(Hypercalcemia WHOIFPI by no disease)}$$
$$= \text{Pr(no disease WHOIFPI)} \cdot r_{\text{no disease} \rightarrow \text{hypercalcemia}}$$
$$= 0.997 \cdot 0.0014 \approx 0.0014$$

The probability that hypercalcemia would have occurred in the first place in the individual can thus be calculated as:

$$\text{Pr(hypercalcemia WHOIFPI)}$$
$$= \text{Pr(hypercalcemia WHOIFPI by PH)} + \text{Pr(hypercalcemia WHOIFPI by cancer)}$$
$$+ \text{Pr(hypercalcemia WHOIFPI by other conditions)} + \text{Pr(hypercalcemia WHOIFPI by no diseas}$$
$$= 0.00125 + 0.0002 + 0.0005 + 0.0014 = 0.00335$$

Subsequently, the probability that hypercalcemia is caused by primary hyperparathyroidism (PH) in the individual can be calculated as:

$$\text{Pr(hypercalcemia is caused by PH in individual)}$$
$$= \frac{\text{Pr(hypercalcemia WHOIFPI by PH)}}{\text{Pr(hypercalcemia WHOIFPI)}}$$
$$= \frac{0.00125}{0.00335} = 0.373 = 37.3\%$$

Similarly, the probability that hypercalcemia is caused by cancer in the individual can be calculated as:

$$\text{Pr(hypercalcemia is caused by cancer in individual)}$$
$$= \frac{\text{Pr(hypercalcemia WHOIFPI by cancer)}}{\text{Pr(hypercalcemia WHOIFPI)}}$$
$$= \frac{0.0002}{0.00335} = 0.060 = 6.0\%,$$

and for other candidate conditions:

$$\text{Pr(hypercalcemia is caused by other conditions in individual)}$$

$$= \frac{\text{Pr(hypercalcemia WHOIFPI by other conditions)}}{\text{Pr(hypercalcemia WHOIFPI)}}$$

$$= \frac{0.0005}{0.00335} = 0.149 = 14.9\%,$$

and the probability that there actually is no disease:

$$\text{Pr(hypercalcemia is present despite no disease in individual)}$$

$$= \frac{\text{Pr(hypercalcemia WHOIFPI by no disease)}}{\text{Pr(hypercalcemia WHOIFPI)}}$$

$$= \frac{0.0014}{0.00335} = 0.418 = 41.8\%$$

For clarification, these calculations are given as the table in the method description:

	PH	Cancer	Other conditions	No disease
P(Condition in population)	0.000125	0.002	-	0.997
RR_x	10	1	-	-
P(Condition WHOIFPI)	0.00125	0.002	-	-
$r_{\text{Condition} \rightarrow \text{hypercalcemia}}$	1	0.1	-	0.0014
P(hypercalcemia WHOIFPI by condition)	0.00125	0.0002	0.0005	0.0014
P(hypercalcemia WHOIFPI) = 0.00335				
P(hypercalcemia is caused by condition in individual)	37.3%	6.0%	14.9%	41.8%

Thus, this method estimates that the probabilities that the hypercalcemia is caused by primary hyperparathyroidism, cancer, other conditions or no disease at all are 37.3%, 6.0%, 14.9% and 41.8%, respectively, which may be used in estimating further test indications.

This case is continued in the example of the method described in the next section.

Likelihood Ratio-based Method

The procedure of differential diagnosis can become extremely complex when fully taking additional tests and treatments into consideration. One method that is somewhat a tradeoff between being clinically perfect and being relatively simple to calculate is one that uses likelihood ratios to derive subsequent post-test likelihoods.

Theory

The initial likelihoods for each candidate condition can be estimated by various methods, such as:

- By epidemiology as described in previous section.

- By clinic-specific pattern recognition, such as statistically knowing that patients coming into a particular clinic with a particular complaint statistically has a particular likelihood of each candidate condition.

One method of estimating likelihoods even after further tests uses likelihood ratios (which is derived from sensitivities and specificities) as a multiplication factor after each test or procedure. In an ideal world, sensitivities and specificities would be established for all tests for all possible pathological conditions. In reality, however, these parameters may only be established for one of the candidate conditions. Multiplying with likelihood ratios necessitates conversion of likelihoods from probabilities to *odds in favor* (hereafter simply termed "odds") by:

$$odds = \frac{probability}{1 - probability}$$

However, only the candidate conditions with known likelihood ratio need this conversion. After multiplication, conversion back to probability is calculated by:

$$probability = \frac{odds}{odds + 1}$$

The rest of the candidate conditions (for which there is no established likelihood ratio for the test at hand) can, for simplicity, be adjusted by subsequently multiplying all candidate conditions with a common factor to again yield a sum of 100%.

The resulting probabilities are used for estimating the indications for further medical tests, treatments or other actions. If there is an indication for an additional test, and it returns with a result, then the procedure is repeated using the likelihood ratio of the additional test. With updated probabilities for each of the candidate conditions, the indications for further tests, treatments or other actions changes as well, and so the procedure can be repeated until an *end point* where there no longer is any indication for currently performing further actions. Such an end point mainly occurs when one candidate condition becomes so certain that no test can be found that is powerful enough to change the relative probability-profile enough to motivate any change in further actions. Tactics for reaching such an end point with as few tests as possible includes making tests with high specificity for conditions of already outstandingly high-profile-relative probability, because the high likelihood ratio positive for such tests is very high, bringing all less likely conditions to relatively lower probabilities. Alternatively, tests with high sensitivity for competing candidate conditions have a high likelihood ratio negative, potentially bringing the probabilities for competing candidate conditions to negligible levels. If such negligible probabilities are achieved, the physician can rule out these conditions, and continue the differential diagnostic procedure with only the remaining candidate conditions.

Example

This example continues for the same patient as in the example for the epidemiology-based method. As with the previous example of epidemiology-based method, this example case is made to demonstrate how this method is applied, but does not represent a guideline for handling similar real-world cases. Also, the example uses relatively specified numbers, while in reality, there are often just rough estimations. In this example, the probabilities for each candidate condition were established by an epidemiology-based method to be as follows:

	PH	Cancer	Other conditions	No disease
Probability	37.3%	6.0%	14.9%	41.8%

These percentages could also have been established by experience at the particular clinic by knowing that these are the percentages for final diagnosis for people presenting to the clinic with hypercalcemia and having a family history of primary hyperparathyroidism.

The condition of highest profile-relative probability (except "no disease") is primary hyperparathyroidism (PH), but cancer is still of major concern, because if it is the actual causative condition for the hypercalcemia, then the choice of whether to treat or not likely means life or death for the patient, in effect potentially putting the indication at a similar level for further tests for both of these conditions.

Here, let's say that the physician considers the profile-relative probabilities of being of enough concern to indicate sending the patient a call for a doctor's visit, with an additional visit to the medical laboratory for an additional blood test complemented with further analyses, including parathyroid hormone for the suspicion of primary hyperparathyroidism.

For simplicity, let's say that the doctor first receives the blood test (in formulas abbreviated as "BT") result for the parathyroid hormone analysis, and that it showed a parathyroid hormone level that is elevated relatively to what would be expected by the calcium level.

Such a constellation can be estimated to have a sensitivity of approximately 70% and a specificity of approximately 90% for primary hyperparathyroidism. This confers a likelihood ratio positive of 7 for primary hyperparathyroidism.

The probability of primary hyperparathyroidism is now termed $Pre\text{-}BT_{PH}$ because it corresponds to before the blood test (Latin preposition *prae* means before). It was estimated at 37.3%, corresponding to an odds of 0.595. With the likelihood ratio positive of 7 for the blood test, the post-test odds is calculated as:

$$Odds(\text{PostBT}_{PH}) = Odds(\text{PreBT}_{PH}) \cdot LH(BT) = 0.595 \cdot 7 = 4.16,$$

where:

- $Odds(PostBT_{PH})$ is the odds for primary hyperparathyroidism after the blood test for parathyroid hormone

- $Odds(PreBT_{PH}$ is the odds in favor of primary hyperparathyroidism before the blood test for parathyroid hormone

- $LH(BT)$ is the likelihood ratio positive for the blood test for parathyroid hormone

An Odds(PostBT$_{PH}$) of 4.16 is again converted to the corresponding probability by:

$$\Pr(\text{PostBT}_{PH}) = \frac{Odds(\text{PostBT}_{PH})}{Odds(\text{PostBT}_{PH})+1} = \frac{4.16}{4.16+1} = 0.806 = 80.6\%$$

The sum of the probabilities for the rest of the candidate conditions should therefore be:

$$\Pr(\text{PostBT}_{rest}) = 100\% - 80.6\% = 19.4\%$$

Before the blood test for parathyroid hormone, the sum of their probabilities were:

$$\Pr(\text{PreBT}_{rest}) = 6.0\% + 14.9\% + 41.8\% = 62.7\%$$

Therefore, to conform to a sum of 100% for all candidate conditions, each of the other candidates must be multiplied by a correcting factor:

$$\text{Correcting factor} = \frac{\Pr(\text{PostBT}_{rest})}{\Pr(\text{PreBT}_{rest})} = \frac{19.4}{62.7} = 0.309$$

For example, the probability of cancer after the test is calculated as:

$$\Pr(\text{PostBT}_{cancer}) = \Pr(\text{PreBT}_{cancer}) \cdot \text{Correcting factor} = 6.0\% \cdot 0.309 = 1.9\%$$

The probabilities for each candidate conditions before and after the blood test are given in following table:

	PH	Cancer	Other conditions	No disease
P(PreBT)	37.3%	6.0%	14.9%	41.8%
P(PostBT)	80.6%	1.9%	4.6%	12.9%

These "new" percentages, including a profile-relative probability of 80% for primary hyperparathyroidism, underlie any indications for further tests, treatments or other actions. In this case, let's say that the physician continues the plan for the patient to attend a doctor's visit for further checkup, especially focused at primary hyperparathyroidism.

A doctor's visit can, theoretically, be regarded as a series of tests, including both questions in a medical history as well as components of a physical examination, where the post-test probability of a previous test can be used as the pre-test probability of the next. The indications for choosing the next test is dynamically influenced by the results of previous tests.

Let's say that the patient in this example is revealed to have at least some of the symptoms and signs of depression, bone pain, joint pain or constipation of more severerity than what would be expected by the hypercalcemia itself, supporting the suspicion of primary hyperparathyroidism, and let's say that the likelihood ratios for the tests, when multiplied together, roughly results in a product of 6 for primary hyperparathyroidism.

The presence of unspecific pathologic symptoms and signs in the history and examination are often concurrently indicative of cancer as well, and let's say that the tests gave an overall likelihood ratio estimated at 1.5 for cancer. For other conditions, as well as the instance of not having any disease at all, let's say that it's unknown how they are affected by the tests at hand, as often happens in reality. This gives the following results for the history and physical examination (abbreviated as P&E):

	PH	Cancer	Other conditions	No disease
P(PreH&E)	80.6%	1.9%	4.6%	12.9%
Odds(PreH&E)	4.15	0.019	0.048	0.148
Likelihood ratio by H&E	6	1.5	-	-
Odds(PostH&E)	24.9	0.0285	-	-
P(PostH&E)	96.1%	2.8%	-	-
Sum of known P(PostH&E)	98.9%			
Sum of the rest P(PostH&E)	1.1%			
Sum of the rest P(PreH&E)	4.6% + 12.9% = 17.5%			
Correcting factor	1.1% / 17.5% = 0.063			
After correction	-	-	0.3%	0.8%
P(PostH&E)	96.1%	2.8%	0.3%	0.8%

These probabilities after the history and examination may make the physician confident enough to plan the patient for surgery for a parathyroidectomy to resect the affected tissue.

At this point, the probability of "other conditions" is so low that the physician cannot think of any test for them that could make a difference that would be substantial enough to form an indication for such a test, and the physician thereby practically regards "other conditions" as ruled out, in this case not primarily by any specific test for such other conditions that were negative, but rather by the absence of positive tests so far.

For "cancer", the cutoff at which to confidently regard it as ruled out may be more stringent because of severe consequences of missing it, so the physician may consider that at least a histopathologic examination of the resected tissue is indicated.

This case is continued in the example of *Combinations* in corresponding section below.

Coverage of Candidate Conditions

The validity of both the initial estimation of probabilities by epidemiology and further workup by likelihood ratios are dependent of inclusion of candidate conditions that are responsible for as large part as possible of the probability of having developed the condition, and it's clinically important to include those where relatively fast initiation of therapy is most likely to result in greatest benefit. If an important candidate condition is missed, no method of differential diagnosis will supply the correct conclusion. The need to find more candidate conditions for inclusion increases with increasing severity of the presentation itself. For example, if the only presentation is a deviating laboratory parameter and all common harmful underlying conditions have been ruled out, then it may be acceptable to stop finding more candidate conditions, but this would much more likely be unacceptable if the presentation would have been severe pain.

Combinations

If two conditions get high post-test probabilities, especially if the sum of the probabilities for conditions with known likelihood ratios become higher than 100%, then the actual condition is a combination of the two. In such cases, that combined condition can be added to the list of candidate conditions, and the calculations should start over from the beginning.

To continue the example used above, let's say that the history and physical examination was indicative of cancer as well, with a likelihood ratio of 3, giving an Odds(PostH&E) of 0.057, corresponding to a P(PostH&E) of 5.4%. This would correspond to a "Sum of known P(PostH&E)" of 101.5%. This is an indication for considering a combination of primary hyperparathyroidism and cancer, such as, in this case, a parathyroid hormone-producing parathyroid carcinoma. A recalculation may therefore be needed, with the first two conditions being separated into "primary hyperparathyroidism without cancer", "cancer without primary hyperparathyroidism" as well as "combined primary hyperparathyroidism and cancer", and likelihood ratios being applied to each condition separately. In this case, however, tissue has already been resected, wherein a histopathologic examination can be performed that includes the possibility of parathyroid carcinoma in the examination (which may entail appropriate sample staining). Let's say that the histopathologic examination confirms primary hyperparathyroidism, but also showed a malignant pattern. By an initial method by epidemiology, the incidence of parathyroid carcinoma is estimated at about 1 in 6 million people per year, giving a very low probability before taking any tests into consideration. In comparison, the probability that a non-malignant primary hyperparathyroidism would have occurred at the same time as an unrelated non-carcinoma cancer that presents with malignant cells in the parathyroid gland is calculated by multiplying the probabilities of the two. The resultant probability is, however, much smaller than the 1 in 6 million. Therefore, the probability of parathyroid carcinoma may still be close to 100% after histopathologic examination despite the low probability of occurring in the first place.

Let's finally say that the diagnosis of parathyroid carcinoma resulted in an extended surgery that removed remaining malignant tissue before it had metastasized, and the patient lived happily ever after.

Machine Differential Diagnosis

Machine differential diagnosis is the use of computer software to partly or fully make a differential diagnosis. It may be regarded as an application of artificial intelligence.

Many studies demonstrate improvement of quality of care and reduction of medical errors by using such decision support systems. Some of these systems are designed for a specific medical problem such as schizophrenia, Lyme disease or ventilator-associated pneumonia. Others such as ESAGIL, Iliad, QMR, DiagnosisPro, VisualDx, Isabel, ZeroMD, DxMate, and Physician Cognition are designed to cover all major clinical and diagnostic findings to assist physicians with faster and more accurate diagnosis.

However, these tools all still require advanced medical skills to rate symptoms and choose additional tests to deduce the probabilities of different diagnoses. Thus, non-professionals should still see a health care provider for a proper diagnosis.

History

The method of differential diagnosis was first suggested for use in the diagnosis of mental disorders by Emil Kraepelin. It is more systematic than the old-fashioned method of diagnosis by *gestalt* (impression).

Alternative Medical Meanings

'Differential diagnosis' is also used more loosely, to refer simply to a list of the most common causes of a given symptom, to a list of disorders similar to a given disorder, or to such lists when they are annotated with advice on how to narrow the list down (the book 'French's Index of Differential Diagnosis', ISBN 0-340-81047-5, is an example). Thus, a differential diagnosis in this sense is medical information specially organized to aid in diagnosis.

Usage Apart from in Medicine

Methods similar to those of differential diagnostic processes in medicine are also is used by biological taxonomists to identify and classify organisms, living and extinct. For example, after finding an unknown species, there can first be a listing of all potential species, followed by ruling out of one by one until, optimally, only one potential choice remains. Similar procedures are used by plant and maintenance engineers, automotive mechanics (less now than previously), and used to be used in diagnosing faulty electronic circuitry. Increasingly often a process of replacement is followed, generally starting with the most inexpensive subunit, rather than the most likely one, until the problem symptoms go away; this is faster if replacement parts are available, and does not require diagnostic skills.

Therapy

Therapy (often abbreviated tx, Tx, or T_x) is the attempted remediation of a health problem, usually following a diagnosis. In the medical field, it is usually synonymous with treatment (also abbreviated tx or T_x). Among psychologists and other mental health professionals, including psychiatrists, psychiatric nurse practitioners, and clinical social workers, the term may refer specifically to psychotherapy (sometimes dubbed 'talking therapy').

As a rule, each therapy has indications and contraindications.

Semantic Field

The words *care*, *therapy*, *treatment*, and *intervention* overlap in a semantic field, and thus they can be synonymous depending on context. Moving rightward through that order, the connotative level of holism decreases and the level of specificity (to concrete instances) increases. Thus, in health care contexts (where its senses are always noncount), the word *care* tends to imply a broad idea of everything done to protect or improve someone's health (for example, as in the terms *preventive care* and *primary care*, which connote ongoing action), although it sometimes implies a narrower idea (for example, in the simplest cases of wound care or postanesthesia care, a few particular steps are sufficient, and the patient's interaction with that provider is soon finished). In contrast, the word *intervention* tends to be specific and concrete, and thus the word is often countable; for example, one instance of cardiac catheterization is one intervention performed, and coronary care (noncount) can require a series of interventions (count). At the extreme, the piling on of such countable interventions amounts to interventionism, a flawed model of care lacking holistic circumspection— merely treating discrete problems (in billable increments) rather than maintaining health. *Therapy* and *treatment*, in the middle of the semantic field, can connote either the holism of *care* or the discreteness of *intervention*, with context conveying the intent in each use. Accordingly, they can be used in both noncount and count senses (for example, *therapy for chronic kidney disease can involve several dialysis treatments per week*).

The words *aceology* and *iamatology* are obscure and obsolete synonyms referring to the study of therapies.

Types of Therapies

Levels of care

Levels of care classify health care into categories of chronology, priority, or intensity, as follows:

- Emergency care handles medical emergencies and is a first point of contact or intake for less serious problems, which can be referred to other levels of care as appropriate.

- Intensive care, also called critical care, is care for extremely ill or injured patients. It thus requires high resource intensity, knowledge, and skill, as well as quick decision making.

- Ambulatory care is care provided on an outpatient basis. Typically patients can walk into and out of the clinic under their own power (hence "ambulatory"), usually on the same day.

- Home care is care at home, including care from providers (such as physicians, nurses, and home health aides) making house calls, care from caregivers such as family members, and patient self-care.

- Primary care is meant to be the main kind of care in general, and ideally a medical home that unifies care across referred providers.

- Secondary care is care provided by medical specialists and other health professionals who generally do not have first contact with patients, for example, cardiologists, urologists and dermatologists. A patient reaches secondary care as a next step from primary care, typically by provider referral although sometimes by patient self-initiative.

- Tertiary care is specialized consultative care, usually for inpatients and on referral from a primary or secondary health professional, in a facility that has personnel and facilities for advanced medical investigation and treatment, such as a tertiary referral hospital.

- Follow-up care is additional care during or after convalescence. Aftercare is generally synonymous with follow-up care.

- End-of-life care is care near the end of one's life. It often includes the following:

    o Palliative care is supportive care, most especially (but not necessarily) near the end of life.

    o Hospice care is palliative care very near the end of life when cure is very unlikely. Its main goal is comfort, both physical and mental.

Lines of Therapy

Treatment decisions often follow formal or informal algorithmic guidelines. Treatment options can often be ranked or prioritized into lines of therapy: first-line therapy, second-line therapy, third-line therapy, and so on. First-line therapy (sometimes called induction therapy, primary therapy, or front-line therapy) is the first therapy that will be tried. Its priority over other options is usually either (1) formally recommended on the basis of clinical trial evidence for its best-available combination of efficacy, safety, and tolerability or (2) chosen based on the clinical experience of the physician. If a first-line therapy either fails to resolve the issue or produces intolerable side effects, additional (second-line) therapies may be substituted or added to the treatment regimen, followed by third-line therapies, and so on.

An example of a context in which the formalization of treatment algorithms and the ranking of lines of therapy is very extensive is chemotherapy regimens. Because of the great difficulty in successfully treating some forms of cancer, one line after another may be tried. In oncology the count of therapy lines may reach 10 or even 20.

Often multiple therapies may be tried simultaneously (combination therapy or polytherapy). Thus combination chemotherapy is also called polychemotherapy, whereas chemotherapy with one agent at a time is called single-agent therapy or monotherapy.

Adjuvant therapy is therapy given in addition to the primary, main, or initial treatment, but simultaneously (as opposed to second-line therapy). Neoadjuvant therapy is therapy that is begun before the main therapy. Thus one can consider surgical excision of a tumor as the first-line therapy for a certain type and stage of cancer even though radiotherapy is used before it; the radiotherapy is neoadjuvant (chronologically first but not primary in the sense of the main event).

Step therapy or stepladder therapy is a specific type of prioritization by lines of therapy.

It is controversial in American health care because unlike conventional decision-making about what constitutes first-line, second-line, and third-line therapy, which in the U.S. reflects safety and efficacy first and cost only according to the patient's wishes, step therapy attempts to mix cost containment by someone other than the patient (third-party payers) into the algorithm. Therapy freedom and the negotiation between individual and group rights are involved.

By Treatment Intent

Therapy type	Description
abortive therapy	A therapy that is intended to stop a medical condition from progressing any further. A medication taken at the earliest signs of a disease, such as an analgesic taken at the very first symptoms of a migraine headache to prevent it from getting worse, is an abortive therapy. Compare abortifacients, which abort a pregnancy.
consolidation therapy	A therapy given to consolidate the gains from induction therapy. In cancer, this means chasing after any malignant cells that may be left.
curative therapy	A therapy with *curative intent*, that is, one that seeks to cure the root cause of a disorder.
definitive therapy	A therapy that may be final, superior to others, curative, or all of those.
destination therapy	A therapy that is the final destination rather than a bridge to another therapy. Usually refers to ventricular assist devices to keep the existing heart going, not just until a heart transplant can occur, but for the rest of the patient's life expectancy.
empiric therapy	A therapy given on an empiric basis; that is, one given according to a clinician's educated guess despite uncertainty about the illness's causative factors. For example, empiric antibiotic therapy administers a broad-spectrum antibiotic immediately on the basis of a good chance (given the history, physical examination findings, and risk factors present) that the illness is bacterial and will respond to that drug (even though the bacterial species or variant is not yet known).
gold standard therapy	A therapy that is definitive, just as a gold standard diagnostic test is a definitive test.
investigational therapy	An experimental therapy. Use of experimental therapies must be ethically justified, because by definition they raise the question of standard of care. Physicians have autonomy to provide empirical care (such as off-label care) according to their experience and clinical judgment, but the autonomy has limits that preclude quackery. Thus it may be necessary to design a clinical trial around the new therapy and to use the therapy only per a formal protocol. Sometimes shorthand phrases such as "treated on protocol" imply not just "treated according to a plan" but specifically "treated with investigational therapy".
maintenance therapy	A therapy taken during disease remission to prevent relapse.
palliative therapy	See supportive therapy for connotative distinctions.
preventive therapy (prophylactic therapy)	A therapy that is intended to prevent a medical condition from occurring (also called prophylaxis). For example, many vaccines prevent infectious diseases.

salvage therapy (rescue therapy)	A therapy tried after others have failed; it may be a "last-line" therapy.
stepdown therapy	Therapy that tapers the dosage gradually rather than abruptly cutting it off. For example, a switch from intravenous to oral antibiotics as an infection is brought under control steps down the intensity of therapy.
supportive therapy	A therapy that does not treat or improve the underlying condition, but instead increases the patient's comfort. For example, supportive care for flu, colds, or gastrointestinal upset can include rest, fluids, and OTC pain relievers; those things don't treat the cause, but they do *treat the symptoms* and thus provide relief. Supportive therapy may be *palliative therapy* (palliative care). The two terms are sometimes synonymous, but *palliative care* often connotes serious illness and end-of-life care, whereas *supportive care* is always connotatively neutral (it may be as simple as mere bedrest for the common cold). Therapy may be categorized as having curative intent (when it is possible to eliminate the disease) or *palliative intent* (when eliminating the disease is impossible and the focus shifts to minimizing the distress that it causes). The two are often contradistinguished (mutually exclusive) in some contexts (such as the management of some cancers), but they are not inherently mutually exclusive; often a therapy can be both curative and palliative simultaneously. Supportive psychotherapy aims to support the patient by alleviating the worst of the symptoms, with the expectation that definitive therapy can follow later if possible.
systemic therapy	A therapy that is systemic. In the physiological sense, this means affecting the whole body (rather than being local or locoregional), whether via systemic administration, systemic effect, or both. Systemic therapy in the psychotherapeutic sense seeks to address people not only on the individual level but also as people in relationships, dealing with the interactions of groups.

By Therapy Composition

Treatments can be classified according to the method of treatment:

By Matter

- by drugs: pharmacotherapy, chemotherapy (also, *medical therapy* often means specifically pharmacotherapy)

- by medical devices: implantation

    o cardiac resynchronization therapy

- by specific molecules: molecular therapy (although most drugs are specific molecules, *molecular medicine* refers in particular to medicine relying on molecular biology)

    o by specific biomolecular targets: targeted therapy

- molecular chaperone therapy
    - by chelation: chelation therapy
- by specific chemical elements:
    - by metals:
        - by heavy metals:
            - by gold: chrysotherapy (aurotherapy)
            - by platinum-containing drugs: platin therapy
        - by biometals
            - by lithium: lithium therapy
            - by potassium: potassium supplementation
            - by magnesium: magnesium supplementation
            - by chromium: chromium supplementation; phonemic neurological hypochromium therapy
            - by copper: copper supplementation
    - by nonmetals:
        - by diatomic oxygen: oxygen therapy, hyperbaric oxygen therapy (hyperbaric medicine)
            - transdermal continuous oxygen therapy
        - by triatomic oxygen (ozone): ozone therapy
        - by fluoride: fluoride therapy
        - by other gases: medical gas therapy
- by water:
    - hydrotherapy
    - aquatic therapy
    - rehydration therapy
        - oral rehydration therapy
    - water cure (therapy)

- by biological materials (biogenic substances, biomolecules, biotic materials, natural products), including their synthetic equivalents: biotherapy

  o by whole organisms

    - by viruses: virotherapy

    - by bacteriophages: phage therapy

    - by animal interaction

  o by constituents or products of organisms

    - by plant parts or extracts (but many drugs are derived from plants, even when the term *phytotherapy* is not used)

      - scientific type: phytotherapy

      - traditional (prescientific) type: herbalism

    - by animal parts: quackery involving shark fins, tiger parts, and so on, often driving threat or endangerment of species

    - by genes: gene therapy

      - gene therapy for epilepsy

      - gene therapy for osteoarthritis

      - gene therapy for color blindness

      - gene therapy of the human retina

      - gene therapy in Parkinson's disease

    - by epigenetics: epigenetic therapy

    - by proteins: protein therapy (but many drugs are proteins despite not being called protein therapy)

    - by enzymes: enzyme replacement therapy

    - by hormones: hormone therapy

      - hormonal therapy (oncology)

      - hormone replacement therapy

        - estrogen replacement therapy

        - androgen replacement therapy

- hormone replacement therapy (menopause)

- hormone replacement therapy (transgender)

  - hormone replacement therapy (male-to-female)

  - hormone replacement therapy (female-to-male)

- antihormone therapy

  - androgen deprivation therapy

- by whole cells: cell therapy (cytotherapy)

  - by stem cells: stem cell therapy

  - by immune cells

- by immune system products: immunotherapy, host modulatory therapy

  - by immune cells:

    - T-cell vaccination

    - cell transfer therapy

    - autologous immune enhancement therapy

    - TK cell therapy

  - by humoral immune factors: antibody therapy

    - by whole serum: serotherapy, including antiserum therapy

    - by immunoglobulins: immunoglobulin therapy

      - by monoclonal antibodies: monoclonal antibody therapy

o by urine: urine therapy (some scientific forms; many prescientific or pseudoscientific forms)

o by food and dietary choices:

- medical nutrition therapy

- grape therapy (quackery)

- by salts (but many drugs are the salts of organic acids, even when drug therapy is not called by names reflecting that)

    o by salts in the air

        - by natural dry salt air: "taking the cure" in desert locales (especially common in prescientific medicine; for example, one 19th-century way to treat tuberculosis)

        - by artificial dry salt air:

            - low-humidity forms of speleotherapy

            - negative air ionization therapy

        - by moist salt air:

            - by natural moist salt air: seaside cure (especially common in prescientific medicine)

            - by artificial moist salt air: water vapor forms of speleotherapy

    o by salts in the water

        - by mineral water: spa cure ("taking the waters") (especially common in prescientific medicine)

        - by seawater: seaside cure (especially common in prescientific medicine)

- by aroma: aromatherapy

- by other materials with mechanism of action unknown

    o by occlusion with duct tape: duct tape occlusion therapy

By Energy

- by electric energy as electric current: electrotherapy, electroconvulsive therapy

- by magnetic energy:

    o magnet therapy

    o pulsed electromagnetic field therapy

    o magnetic resonance therapy

- by electromagnetic radiation (EMR):
  - by light: light therapy (phototherapy)
    - ultraviolet light therapy
      - PUVA therapy
    - photodynamic therapy
      - photothermal therapy
      - cytoluminescent therapy
    - blood irradiation therapy
    - by darkness: dark therapy
    - by lasers: laser therapy
      - low level laser therapy
  - by gamma rays: radiosurgery
    - Gamma Knife radiosurgery
    - stereotactic radiation therapy
    - cobalt therapy
  - by radiation generally: radiation therapy (radiotherapy)
    - intraoperative radiation therapy
    - by EMR particles:
      - particle therapy
        - proton therapy
        - electron therapy
          - intraoperative electron radiation therapy
          - Auger therapy
        - neutron therapy
          - fast neutron therapy
          - neutron capture therapy of cancer

- by radioisotopes emitting EMR:
    - by nuclear medicine
    - by brachytherapy
  - quackery type: electromagnetic therapy (alternative medicine)
- by mechanical: manual therapy as massotherapy and therapy by exercise as in physiotherapy and exercise therapy
  - inversion therapy
- by sound:
  - by ultrasound:
    - ultrasonic lithotripsy
      - extracorporeal shock wave lithotripsy
      - extracorporeal shockwave therapy
    - sonodynamic therapy (largely pseudoscientific)
  - by music: music therapy
    - neurologic music therapy
- by temperature
  - by heat: heat therapy (thermotherapy)
    - by moderately elevated ambient temperatures: hyperthermia therapy
      - by dry warm surroundings: Waon therapy
      - by dry or humid warm surroundings: sauna, including infrared sauna, for sweat therapy
  - by cold:
    - by extreme cold to specific tissue volumes: cryotherapy
    - by ice and compression: cold compression therapy
    - by ambient cold: hypothermia therapy for neonatal encephalopathy
  - by hot and cold alternation: contrast bath therapy

By Human Interaction

- by counseling, such as psychotherapy
    - systemic therapy
    - by group psychotherapy
- by cognitive behavioral therapy
    - by cognitive therapy
    - by cognitive rehabilitation therapy
    - by behaviour therapy
        - by dialectical behavior therapy
    - by cognitive emotional behavioral therapy
- by family therapy
- by education
    - by psychoeducation
    - by information therapy
- by physical therapy/occupational therapy, massage therapy, chiropractic or acupuncture
- by lifestyle modifications, such as avoiding unhealthy food or maintaining a predictable sleep schedule
- by coaching

By Animal Interaction

- by pets, assistance animals, or working animals: animal-assisted therapy
    - by horses: equine therapy, hippotherapy
    - by dogs: pet therapy with therapy dogs, including grief therapy dogs
    - by cats: pet therapy with therapy cats, including Oscar
- by fish: ichthyotherapy (wading with fish), aquarium therapy (watching fish)
- by maggots: maggot therapy
- by worms:

- o   by internal worms: helminthic therapy
- o   by leechs: leech therapy

By Meditation

- by mindfulness: mindfulness-based cognitive therapy

By Reading

- by bibliotherapy

By Creativity

- by expression: expressive therapy
  - o   by writing: writing therapy
    - ▪ journal therapy
- by play: play therapy
- by art: art therapy
  - o   sensory art therapy
  - o   comic book therapy
- by gardening: horticultural therapy
- by dance: dance therapy
- by drama: drama therapy
- by recreation: recreational therapy
- by music: music therapy

By Sleeping and Waking

- by deep sleep: deep sleep therapy
- by waking: wake therapy

2009 Flu Pandemic Vaccine

The 2009 flu pandemic vaccines are the set of influenza vaccines that have been developed to protect against the pandemic H1N1/09 virus. These vaccines either contain inactivated (killed) influenza virus, or weakened live virus that cannot cause influenza. The killed vaccine is injected, while the live vaccine is given as an interperineal nasal spray. Both these types of vaccine are usually produced by growing the virus in chicken eggs. Around three billion doses will be produced annually, with delivery from November 2009.

In studies, the vaccine appears both effective and safe, providing a strong protective immune response and having similar safety profile to the normal seasonal influenza vaccine. However, about 30% of people already have some immunity to the virus, with the vaccine conferring greatest benefit on young people, since many older people are already immune through exposure to similar viruses in the past. The vaccine also provides some cross-protection against the 1918 flu pandemic strain.

Early results (pre-25 December 2009) from an observational cohort of 248,000 individuals in Scotland have shown the vaccine to be effective at preventing H1N1 influenza (95.0% effectiveness [95% confidence intervals (CI) 76.0–100.0]) and influenza related hospital admissions (64.7% [95%CI 12.0–85.8]).

Developing, testing, and manufacturing sufficient quantities of a vaccine is a process that takes many months. According to Keiji Fukuda of the World Health Organization (WHO), "There's much greater vaccine capacity than there was a few years ago, but there is not enough vaccine capacity to instantly make vaccines for the entire world's population for influenza." Nasal mist version of the vaccine started shipping on 1 October 2009.

Types of Vaccine

Two types of influenza vaccines are available:

- TIV (flu shot (injection) of trivalent (three strains; usually A/H1N1, A/H3N2, and B) inactivated (killed) vaccine) or

- LAIV (nasal spray (mist) of live attenuated influenza vaccine.)

TIV works by putting into the bloodstream those parts of three strains of flu virus that the body uses to create antibodies; while LAIV works by inoculating the body with those same three strains, but in a modified form that cannot cause illness.

LAIV is not recommended for individuals under age 2 or over age 49, but might be comparatively more effective among children over age two.

Manufacturing Methods

For the inactivated vaccines, the virus is grown by injecting it, along with some antibiotics, into fertilized chicken eggs. About one to two eggs are needed to make each dose of vaccine. The virus replicates within the allantois of the embryo, which is the equivalent of the placenta in mammals. The fluid in this structure is removed and the virus purified from this fluid by methods such as filtration or centrifugation. The purified viruses are then inactivated ("killed") with a small amount of a disinfectant. The inactivated virus is treated with detergent to break up the virus into particles, and the broken capsule segments and released proteins are concentrated by centrifugation. The final preparation is suspended in sterile phosphate buffered saline ready for injection. This vaccine mainly contains the killed virus but might also contain tiny amounts of egg protein and the antibiotics, disinfectant and detergent used in the manufacturing process. In multi-dose versions of the vaccine, the preservative thimerosal is added to prevent growth of bacteria. In some versions of the vaccine used in Europe and Canada, such as *Arepanrix* and *Fluad*, an adjuvant is also added, this contains a fish oil called squalene, vitamin E and an emulsifier called polysorbate 80.

For the live vaccine, the virus is first adapted to grow at 25 °C (77 °F) and then grown at this temperature until it loses the ability to cause illness in humans, which would require the virus to grow at our normal body temperature of 37 °C (99 °F). Multiple mutations are needed for the virus to grow at cold temperatures, so this process is effectively irreversible and once the virus has lost virulence (become "attenuated"), it will not regain the ability to infect people. To make the vaccine, the attenuated virus is grown in chicken eggs as before. The virus-containing fluid is harvested and the virus purified by filtration; this step also removes any contaminating bacteria. The filtered preparation is then diluted into a solution that stabilizes the virus. This solution contains monosodium glutamate, potassium phosphate, gelatin, the antibiotic gentamicin, and sugar.

A new method of producing influenza virus is used to produce the Novartis vaccine Optaflu. In this vaccine the virus is grown in cell culture instead of in eggs. This method is faster than the classic egg-based system and produces a purer final product. Importantly, there are no traces of egg proteins in the final product, so the vaccine is safe for people with egg allergies.

Previous Seasonal Vaccine Production

The WHO recommended before the H1N1/09 outbreak that vaccines for the Northern Hemisphere's 2009–2010 flu season contain an A(H1N1)-like virus, and stocks have been made. However, the strain of H1N1 in the seasonal flu vaccine is different from the new pandemic strain H1N1/09 and offers no immunity against it. The US Centers for Disease Control and Prevention (CDC) characterized over 80 new H1N1 viruses that may be used in a vaccine.

Production Questions and Decisions

Questions

There was concern in mid-2009 that, should a second, deadlier wave of this new H1N1 strain appear during the northern autumn of 2009, producing pandemic vaccines ahead of time could turn out to be a serious waste of resources as the vaccine might not be effective against it, and there would also be a shortage of seasonal flu vaccine available if production facilities were switched to the new vaccine. Seasonal flu vaccine was being made as of May 2009, according to WebMD. The news site added that although vaccine makers would be ready to switch to making a swine flu vaccine, many questions remained unanswered, including: "Should we really make a swine flu vaccine? Should we base a vaccine on the current virus, since flu viruses change rapidly? Vaccine against the current virus might be far less effective against a changed virus – should we wait to see if the virus changes? If vaccine production doesn't start soon, swine flu vaccine won't be ready when it's needed."

The costs of producing a vaccine also became an issue, with some U.S. lawmakers questioning whether a new vaccine was worth the unknown benefits. Representatives Phil Gingrey and Paul Broun, for instance, were not convinced that the U.S. should spend up to US$2 billion to produce one, with Gingrey stating "We can't let all of our spending and our reaction be media-driven in responding to a panic so that we don't get Katrina-ed. ... It's important because what we are talking about as we discuss the appropriateness of spending $2 billion to produce a vaccine that may never be used – that is a very important decision that our country has to make." In fact, a Fairleigh Dickinson University PublicMind poll found in October 2009 that a majority (62%) of New Jerseyans were not planning on getting the vaccine at all.

Before the pandemic was declared, the WHO said that if a pandemic was declared it would attempt to make sure that a substantial amount of vaccine was available for the benefit of developing countries. Vaccine makers and countries with standing orders, such as the U.S. and a number of European countries, would be asked, according to WHO officials, "to share with developing countries from the moment the first batches are ready if an H1N1 vaccine is made" for a pandemic strain. The global body stated that it wanted companies to donate at least 10% of their production or offer reduced prices for poor countries that could otherwise be left without vaccines if there is a sudden surge in demand.

Gennady Onishchenko, Russia's chief doctor, said on 2 June 2009 that swine flu was not aggressive enough to cause a worldwide pandemic, noting that the current mortality rate of confirmed cases was 1.6% in Mexico and only 0.1% in the United States. He stated at a press conference, "So far it is unclear if we need to use vaccines against the flu because the virus that is now circulating throughout Europe and North America does not have a pandemic nature." In his opinion, a vaccine could be produced, but said that preparing a vaccine now would be considered "practice," since the world would soon need a new vaccine against a new virus. "What's 16,000 sick people? During any flu season, some 10,000 a day become ill in Moscow alone," he said.

Production Timelines

After a meeting with the WHO on 14 May 2009, pharmaceutical companies said they were ready to begin making a swine flu vaccine. According to news reports, the WHO's experts will present recommendations to WHO Director-General Margaret Chan, who was expected to issue advice to vaccine manufacturers and the Sixty-second World Health Assembly. WHO's Keiji Fukuda told reporters "These are enormously complicated questions, and they are not something that anyone can make in a single meeting." Most flu vaccine companies can not make both seasonal flu vaccine and pandemic flu vaccine at the same time. Production takes months and it is impossible to switch halfway through if health officials make a mistake. If the swine flu mutates, scientists aren't sure how effective a vaccine made now from the current strain will remain. Rather than wait on the WHO decision, however, some countries in Europe have decided to go ahead with early vaccine orders.

On 20 May 2009, AP reported: "Manufacturers won't be able to start making the [swine flu] vaccine until mid-July at the earliest, weeks later than previous predictions, according to an expert panel convened by WHO. It will then take months to produce the vaccine in large quantities. The swine flu virus is not growing very fast in laboratories, making it difficult for scientists to get the key ingredient they need for a vaccine, the 'seed stock' from the virus [...] In any case, mass producing a pandemic vaccine would be a gamble, as it would take away manufacturing capacity for the seasonal flu vaccine for the flu that kills up to 500,000 people each year. Some experts have wondered whether the world really needs a vaccine for an illness that so far appears mild."

Another option proposed by the CDC was an "earlier rollout of seasonal vaccine," according to the CDC's Daniel Jernigan. He said the CDC would work with vaccine manufacturers and experts to see if that would be possible and desirable. Flu vaccination usually starts in September in the United States and peaks in November. Some vaccine experts agree it would be better to launch a second round of vaccinations against the new H1N1 strain instead of trying to add it to the seasonal flu vaccine or replacing one of its three components with the new H1N1 virus.

The Australian company CSL said that they were developing a vaccine for the swine flu and predicted that a suitable vaccine would be ready by August. However, John Sterling, Editor in Chief of *Genetic Engineering & Biotechnology News*, said on 2 June, "It can take five or six months to come up with an entirely novel influenza vaccine. There is a great deal of hope that biotech and pharma companies might be able to have something ready sooner."

As of September 2009 a vaccine for H1N1/09 was expected to be available starting in November 2009, with production of three billion doses per year. It was expected that two doses would be needed to provide sufficient protection, but tests indicated that one dose would be sufficient for adults.

As of 28 September 2009 GlaxoSmithKline produced a vaccine made by growing the virus in hens' eggs, then breaking and deactivating the virus, and Baxter International produced a vaccine made in cell culture, suitable for those who have an egg allergy. The vaccines have been approved for use in the European Union.

Testing

Initial Phase I human testing began with Novartis' MF59 candidate in July 2009, at which time phase II trials of CSL's candidate CSL425 vaccine were planned to start in August 2009, but had not begun recruiting. Sanofi Pasteur's candidate inactivated H1N1 had several phase II trials planned as of 21 July 2009, but had not begun recruiting. News coverage conflicted with this information, as Australian trials of the CSL candidate were announced as having started on 21 July, and the Chinese government announced the start of trials of the Hualan Biological Engineering candidate.

Pandemrix, made by GlaxoSmithKline (GSK), and Focetria, made by Novartis were approved by the European Medicines Agency on 25 September 2009, and Celvapan, made by Baxter was approved the following week. The first comparative clinical study of both vaccines started on children in the United Kingdom on 25 September 2009. GSK announced results from clinical trials assessing the use of Pandemrix in children, adults, and the elderly. A 2009 trial examined the safety and efficacy of two different doses of the split-virus vaccine, and was published in *The New England Journal of Medicine*. The vaccine used in the trail was prepared by CSL Biotherapies in chicken eggs, in the same way as the seasonal vaccine. A robust immune response was produced in over 90% of patients after a single dose of either 15 or 30 µg of antigen. This study suggested that the current recommendation for two doses of vaccine are overkill and that a single dose is quite sufficient.

Arepanrix, an AS03-Adjuvanted H1N1 Pandemic Influenza Vaccine similar to Pandemrix and also made by GSK, was authorized by Canada's Minister of Health on 21 October 2009.

Adverse Events

A review by the U.S. National Institutes of Health (NIH) concluded that the 2009 H1N1 ("swine flu") vaccine has a safety profile similar to that of seasonal vaccine.

In an initial clinical trial in Australia, non-serious adverse events were reported by about half of the 240 people vaccinated, with these events including tenderness and pain at the site of injection, headache, malaise, and muscle pain. Two people had more severe events, with a much longer spell of nausea, muscle pain and malaise that lasted several days. The authors stated that the frequency and severity of these adverse events were similar to those normally seen with seasonal influenza vaccines. A second trial involved 2,200 people ranging from 3 to 77 years of age. In this study no patients reported serious adverse events, with the most commonly observed events being pain at the injection site and fever, which occurred in 10–25% of people. Although this trial followed up patients individually, the Government has been criticized for relying on voluntary reporting for post-vaccination evaluation in other circumstances, since this is "unlikely to accurately measure the percentage of people who get adverse effect".

As of 19 November 2009, the World Health Organization (WHO) said that 65 million doses of vaccine had been administered and that it had a similar safety profile to the seasonal flu vaccine, with no significant differences in the adverse events produced by the different types of vaccine. There has been one report of an adverse event per 10,000 doses of vaccine, with only five percent of these adverse events being serious, an overall rate of serious events of one in 200,000 doses.

In Canada, after 6.6 million doses of vaccine had been distributed between 21 October and 7 November, there were reports of mild adverse events in 598 people vaccinated including: nausea, dizziness, headache, fever, vomiting, and swelling or soreness at the injection site. There were reports of tingling lips or tongue, difficulty breathing, hives, and skin rashes. Thirty six people had serious adverse events, including anaphylaxis and febrile convulsions. The rate of serious adverse events is one in 200,000 doses distributed, which according to Canada's chief public health officer, is less than expected for the seasonal flu vaccine. GlaxoSmithKline recalled a batch of vaccine in Canada after it appeared to cause higher rates of adverse events than other batches.

In the USA 46 million doses had been distributed as of 20 November 2009 and 3182 adverse events were reported. The CDC stated that the "vast majority" were mild, with about one serious adverse event in 260,000 doses.

In Japan around 15 million people had been vaccinated by 31 December 2009. 1,900 cases of side effects and 104 cases of death were reported from medical institutions. The health ministry announced that it will conduct epidemiologic investigation.

In France, around five million people had been vaccinated by 30 December 2009. 2,657 cases of side effects, eight cases of intrauterine death and five cases of miscarriages were reported after vaccination by afssaps.

Rare potential adverse events are temporary bleeding disorders and Guillain–Barré syndrome (GBS), a serious condition involving the peripheral nervous system, from which most patients recovery fully within a few months to a year. Some studies have indicated that influenza-like illness is itself associated with an increased risk of GBS, suggesting that vaccination might indirectly protect against the disorder by protecting against flu. According to Marie-Paule Kieny of WHO assessing the side-effects of large-scale influenza vaccination is complicated by the fact that in any large population a few people will become ill and die at any time. For example, in any six-week period in the UK six sudden deaths from unknown causes and 22 cases of Guillain–Barré syndrome would be expected, so if everyone in the UK were vaccinated, this background rate of illness and death would continue as normal and some people would die simply by chance soon after the vaccination.

Some scientists have reported concerns about the longer-term effects of the vaccine. For instance, Sucharit Bhakdi, professor of medical microbiology at the Johannes Gutenberg University of Mainz in Germany, wrote in the journal, *Medical Microbiology and Immunology*, of the possibility that immune stimulation by vaccines or any other cause might worsen pre-existing heart disease. Chris Shaw, a neuroscientist at the University of British Columbia, expressed concern that serious side-effects may not appear immediately; he said it took five to ten years to see most of the Gulf War syndrome outcomes.

The CDC states that most studies on modern influenza vaccines have seen no link with GBS, Although one review gives an incidence of about one case per million vaccinations, a large study in China, reported in *The New England Journal of Medicine* covering close to 100 million doses of H1N1 flu vaccine found only eleven cases of Guillain–Barré syndrome, actually lower than the normal rate of the disease in China, and no other notable side effects.

Pregnant Women and Children

A 2009 review of the use of influenza vaccines in pregnant women stated that influenza infections posed a major risk during pregnancy and that multiple studies had shown that the inactivated vaccine was safe in pregnant women, concluding that this vaccine "can be safely and effectively administered during any trimester of pregnancy" and that high levels of immunization would avert "a significant number of deaths". A 2004 review of the safety of influenza vaccines in children stated that the live vaccine had been shown to be safe but that it might trigger wheezing in some children with asthma; less data for the trivalent inactivated vaccine was available, but no serious symptoms had been seen in clinical trials.

Squalene

Newsweek states that "wild rumours" about the swine flu vaccine are being spread through e-mails, it writes that "The claims are nearly pure bunk, with only trace amounts of fact." These rumours generally make unfounded claims that the vaccine is dangerous and they may also promote conspiracy theories. For example, *Newsweek* states that some chain e-mails make false claims about squalene (shark liver oil) in vaccines. *The New York Times* also notes that anti-vaccine groups have spread "dire warnings" about formulations of the vaccine that contain squalene as an adjuvant. An adjuvant is a substance that boosts the body's immune response, thereby stretching the supply of the vaccine and helping immunize elderly people with a weak immune system. Squalene is a normal part of the human body, made in the liver and circulating in the blood, and is also found in many foods, such as eggs and olive oil. None of the formulations of vaccine used in the US contain squalene, or any other adjuvant. However, some European and Canadian formulations do contain 25 µg of squalene per dose, which is roughly the amount found in a drop of olive oil. Some animal experiments have suggested that squalene might be linked to autoimmune disorders. although others suggest squalene might protect people against cancer.

Squalene-based adjuvants have been used in European influenza vaccines since 1997, with about 22 million doses administered over the past twelve years. The WHO states that no severe side effects have been associated with these vaccines, although they can produce mild inflammation at the site of injection. The safety of squalene-containing influenza vaccines have also been tested in two separate clinical trials, one with healthy non-elderly people, and one with elderly people, in both trials the vaccine was safe and well tolerated, with only weak side-effects, such as mild pain at the injection site. A 2009 meta-analysis brought together data from 64 clinical trials of influenza vaccines with the squalene-containing adjuvant MF59 and compared them to the effects of vaccines with no adjuvant. The analysis reported that the adjuvanted vaccines were associated with slightly lower risks of chronic diseases, but that neither type of vaccines altered the normal rate of autoimmune diseases; the authors concluded that their data "supports the good safety profile associated with MF59-adjuvanted influenza vaccines and suggests there may be a clinical benefit over non-MF59-containing vaccines". A 2004 review of the effects of adjuvants on mice and humans concluded that "despite numerous case reports on vaccination induced autoimmunity, most epidemiological studies failed to confirm the association and the risk appears to be extremely low or non-existent", although the authors noted that the possibility that adjuvants might cause damaging immune reactions in a few susceptible people has not been completely ruled out. A 2009 review of oil-based adjuvants in influenza vaccines stated that this type of adjuvant "neither stimulates antibodies against squalene oil naturally produced by the humans body nor enhances titers of preexisting antibodies to squalene" and that these formulations did not raise any safety concerns.

A paper published in 2000 suggested that squalene might have caused of Gulf War syndrome by producing anti-squalene antibodies, although other scientists stated that it was uncertain if the methods used were actually capable of detecting these antibodies. A 2009 U.S. Department of Defense study comparing healthy Navy personnel to those suffering from Gulf War syndrome was published in the journal *Vaccine*, this used a validated test for these antibodies and found no link between the presence of the antibodies and illness, with about half of both groups having these antibodies and no correlation between symptoms and antibodies. Furthermore, none of the vaccines given to US troops during the Gulf war actually contained any squalene adjuvants.

Thiomersal

Multi-dose versions of the vaccine contain the preservative thiomersal (also known as thimerosal), a mercury compound that prevents contamination when the vial is used repeatedly. Single-dose versions and the live vaccine do not contain this preservative. In the U.S., one dose from a multi-dose vial contains approximately 25 micrograms of mercury, a bit less than a typical tuna fish sandwich. In Canada, different variants contain five and 50 micrograms of thimerosal per dose. The use of thiomersal has been controversial, with claims that it can cause autism and other developmental disorders.

The U.S. Institute of Medicine examined these claims and concluded in 2004 that the evidence did not support any link between vaccines and autism. Other reviews came to similar conclusions, with a 2006 review in the *Canadian Journal of Neurological Sciences* stating that there is no convincing evidence to support the claim that thimerosal has a causal role in autism, and a 2009 review in the journal *Clinical Infectious Diseases* stating that claims that mercury can cause autism are "biologically implausible". The U.K. National Health Service stated in 2003 that "There is no evidence of long-term adverse effects due to the exposure levels of thiomersal in vaccines." The World Health Organization concluded that there is "no evidence of toxicity in infants, children or adults exposed to thiomersal in vaccines". Indeed, in 2008 a review noted that even though thiomersal was removed from all US childhood vaccines in 2001, this has not changed the number of autism diagnoses, which are still increasing.

Dystonia

According to the CDC, there is no evidence either for or against dystonia being caused by the vaccinations. Dystonia is extremely rare. Due to the very low numbers of cases, dystonia is poorly understood. There were only five cases noted that might have been associated with influenza vaccinations over a span of eighteen years. In one recent case, a woman noted flu-like symptoms, followed by difficulties with movement and speech starting ten days after a seasonal influenza vaccination. However the Dystonia Medical Research Foundation stated that it is unlikely that the symptoms in this case were actually dystonia and stated that there has "never been a validated case of dystonia resulting from a flu shot".

Children Vaccine Recall

On 15 December 2009, One of the five manufacturers supplying the H1N1 vaccine to the United States recalled thousands of doses because they were not as potent as expected. The French manufacturer Sanofi Pasteur voluntarily recalled about 800,000 doses of vaccine meant for children between the ages of six months and 35 months. The company and the Centers for Disease Control and Prevention (CDC) emphasized that the recall was not prompted by safety concerns, and that even though the vaccine is not quite as potent as it is supposed to be, children who received it do not need to be immunized again. The CDC emphasized that there is no danger for any child who received the recalled vaccine. When asked what parents should do, CDC spokesman Tom Skinner said, "absolutely nothing." He said if children receive this vaccine, they will be fine.

Pandemrix-related Increase of Narcolepsy in Finland and Sweden

In 2010, The Swedish Medical Products Agency (MPA) and The Finnish National Institute for Health and Welfare (THL) received reports from Swedish and Finnish health care professionals regarding narcolepsy as suspected adverse drug reaction following Pandemrix flu vaccination. The reports concern children aged 12–16 years where symptoms compatible with narcolepsy, diagnosed after thorough medical investigation, have occurred one to two months after vaccination.

THL concluded in February 2011 that there is a clear connection between the Pandemrix vaccination campaign of 2009 and 2010 and narcolepsy epidemic in Finland: there was a nine times higher probability to get narcolepsy with vaccination than without it.

At the end of March 2011, an MPA press release stated: "Results from a Swedish registry based cohort study indicate a 4-fold increased risk of narcolepsy in children and adolescents below the age of 20 vaccinated with Pandemrix, compared to children of the same age that were not vaccinated." The same study found no increased risk in adults who were vaccinated with Pandemrix.

Availability

Centers for Disease Control and Prevention

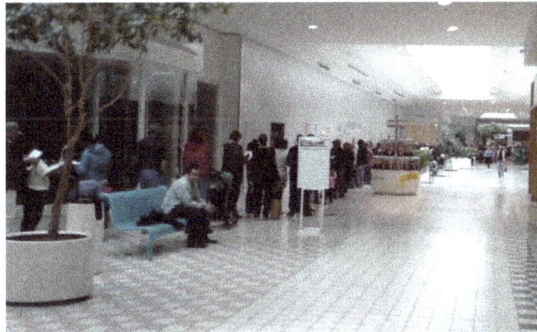

2,500 people line up in a mall in Texas City, Texas to receive the H1N1 vaccine from the Galveston County Health Department on 30 October 2009.

The American Centers for Disease Control and Prevention issued the following recommendations on who should be vaccinated (order is not in priority):

- Pregnant women, because they are at higher risk of complications and can potentially provide protection to infants who cannot be vaccinated;

- Household contacts and caregivers for children younger than 6 months of age, because younger infants are at higher risk of influenza-related complications and cannot be vaccinated. Vaccination of those in close contact with infants younger than 6 months old might help protect infants by "cocooning" them from the virus;

- Healthcare and emergency medical services personnel, because infections among healthcare workers have been reported and this can be a potential source of infection for vulnerable patients. Also, increased absenteeism in this population could reduce healthcare system capacity;

- All people from 6 months through 24 years of age:

    - Children from 6 months through 18 years of age, because cases of 2009 H1N1 influenza have been seen in children who are in close contact with each other in school and day care settings, which increases the likelihood of disease spread, and

    - Young adults 19 through 24 years of age, because many cases of 2009 H1N1 influenza have been seen in these healthy young adults and they often live, work, and study in close proximity, and they are a frequently mobile population; and,

- Persons aged 25 through 64 years who have health conditions associated with higher risk of medical complications from influenza.

- Once the demand for these groups has been met at a local level, everyone from the ages of 25 through 64 years should be vaccinated too.

In addition, the CDC recommends

Children through 9 years of age should get two doses of vaccine, about a month apart. Older children and adults need only one dose.

National Health Service

The UK's National Health Service policy is to provide vaccine in this order of priority:

- People aged between six months and 65 years with:

    - chronic lung disease;

    - chronic heart disease;

    - chronic kidney disease;

    - chronic liver disease;

    - chronic neurological disease;

    - diabetes; or

    - suppressed immune system, whether due to disease or treatment.

- All pregnant women.

- People who live with someone whose immune system is compromised (for example, people with cancer or HIV/AIDS).

- People aged 65 and over in the seasonal flu vaccine at-risk groups.

This excludes the large majority of individuals aged six months to 24 years, a group for which the CDC recommends vaccination.

The NHS notes that:

- Healthy people over 65 years of age seem to have some natural immunity.

- Children, while disproportionately affected, tend to make full recoveries.

- The vaccine is ineffective in young infants.

The United Kingdom began its administration program 21 October 2009. UK Soldiers serving in Afghanistan will also be offered vaccination.

Surplus Vaccine

By April 2010, it was apparent that most of the vaccine was not needed. The US government had bought 229 million doses of H1N1 vaccines of which 91 million doses were used; of the surplus, some of it was stored in bulk, some of it was sent to developing countries and 71 million doses will be destroyed. The World Health Organization is planning to examine if it overreacted to the H1N1 outbreak.

Political Issues

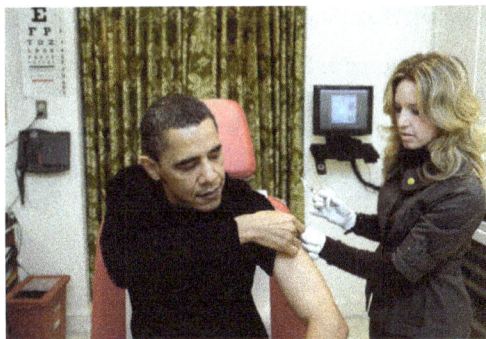

US President Barack Obama receives the vaccine on 20 December 2009

General political issues, not restricted to the 2009 outbreak, arose regarding the distribution of vaccine. In many countries supplies are controlled by national or local governments, and the question of how the vaccine will be allocated should there be an insufficient supply for everyone is critical, and will likely depend on the patterns of any pandemic, and the age groups most at risk for serious complications, including death. In the case of a lethal pandemic people will be demanding access to the vaccine and the major problem will be making it available to those who need it.

While it has been suggested that compulsory vaccination may be needed to control a pandemic, many countries do not have a legal framework that would allow this. The only populations easily compelled to accept vaccination are military personnel (who can be given routine vaccinations as part of their service obligations), health care personnel (who can be required to be vaccinated to protect patients), and school children, who (under United States constitutional law) could be required to be vaccinated as a condition of attending school.

References

- National Academies of Sciences, Engineering, and Medicine (2015). Improving Diagnosis in Health Care. Washington, DC: The National Academies Press. p. S-1. doi:10.17226/21794. ISBN 978-0-309-37769-0.

- Treasure, Wilfrid (2011). "Chapter 1: Diagnosis". Diagnosis and Risk Management in Primary Care: words that count, numbers that speak. Oxford: Radcliffe. ISBN 978-1-84619-477-1.

- Board on Health Promotion and Disease Prevention (HPDP). "Immunization Safety Review: Vaccines and Autism". Institute of Medicine (IOM). National Academies Press. May 2004. ISBN 0-309-09237-X.

- Collignon, P (April 2010). "H1N1 immunisation: Too much, too soon?" (PDF). Aust Prescr. 33 (2): 30–31. ISSN 0312-8008. Retrieved 13 February 2011.

- "WHO seeks swine flu vaccine help for poor nations". USA Today. Associated Press. 20 May 2009. Retrieved 13 February 2011.

- Fox, Maggie (20 May 2009). "U.S. officials consider bumping up flu shot season". Reuters. Archived from the original on 3 February 2011. Retrieved 13 February 2011.

- "Pandemic 2009 Influenza Update: Pandemrix data in children and adolescents from 3 to 17 years of age" (Press release). GlaxoSmithKline. 23 November 2009. Retrieved 13 February 2011.

- Mariner, Wendy K; Annas, George J; Parmet, Wendy E (2 May 2009). "Pandemic Preparedness: A Return to the Rule of Law" (PDF). Drexel Law Review. 1 (2): 341–82. SSRN 1399066. Retrieved 5 February 2011.

5

Major Drawbacks of Biomedicine

The illnesses elaborated in this chapter are HIV, HIV/AIDS and swine influenza. The chapter provides an integrated understanding on the current concerns surrounding biomedical research. Biomedicine has developed a reputation for focusing too much on treatment, rather than prevention, which cannot be very useful when it comes to illnesses like HIV and swine influenza. This chapter on HIV/AIDS and swine influenza offers an insightful focus, keeping in mind the complex subject matter.

HIV

The human immunodeficiency virus (HIV) is a lentivirus (a subgroup of retrovirus) that causes HIV infection and over time acquired immunodeficiency syndrome (AIDS). AIDS is a condition in humans in which progressive failure of the immune system allows life-threatening opportunistic infections and cancers to thrive. Without treatment, average survival time after infection with HIV is estimated to be 9 to 11 years, depending on the HIV subtype. Infection with HIV occurs by the transfer of blood, semen, vaginal fluid, pre-ejaculate, or breast milk. Within these bodily fluids, HIV is present as both free virus particles and virus within infected immune cells.

HIV infects vital cells in the human immune system such as helper T cells (specifically $CD4^+$ T cells), macrophages, and dendritic cells. HIV infection leads to low levels of $CD4^+$ T cells through a number of mechanisms, including pyroptosis of abortively infected T cells, apoptosis of uninfected bystander cells, direct viral killing of infected cells, and killing of infected $CD4^+$ T cells by CD8 cytotoxic lymphocytes that recognize infected cells. When $CD4^+$ T cell numbers decline below a critical level, cell-mediated immunity is lost, and the body becomes progressively more susceptible to opportunistic infections.

Virology

Classification

Comparison of HIV species				
Species	**Virulence**	**Infectivity**	**Prevalence**	**Inferred origin**
HIV-1	High	High	Global	Common chimpanzee
HIV-2	Lower	Low	West Africa	Sooty mangabey

HIV is a member of the genus *Lentivirus*, part of the family *Retroviridae*. Lentiviruses have many morphologies and biological properties in common. Many species are infected by lentiviruses, which are characteristically responsible for long-duration illnesses with a long incubation period. Lentiviruses are transmitted as single-stranded, positive-sense, enveloped RNA viruses. Upon entry into the target cell, the viral RNA genome is converted (reverse transcribed) into double-stranded DNA by a virally encoded reverse transcriptase that is transported along with the viral genome in the virus particle. The resulting viral DNA is then imported into the cell nucleus and integrated into the cellular DNA by a virally encoded integrase and host co-factors. Once integrated, the virus may become latent, allowing the virus and its host cell to avoid detection by the immune system. Alternatively, the virus may be transcribed, producing new RNA genomes and viral proteins that are packaged and released from the cell as new virus particles that begin the replication cycle anew.

Two types of HIV have been characterized: HIV-1 and HIV-2. HIV-1 is the virus that was initially discovered and termed both LAV and HTLV-III. It is more virulent, more infective, and is the cause of the majority of HIV infections globally. The lower infectivity of HIV-2 compared to HIV-1 implies that fewer of those exposed to HIV-2 will be infected per exposure. Because of its relatively poor capacity for transmission, HIV-2 is largely confined to West Africa.

Structure and Genome

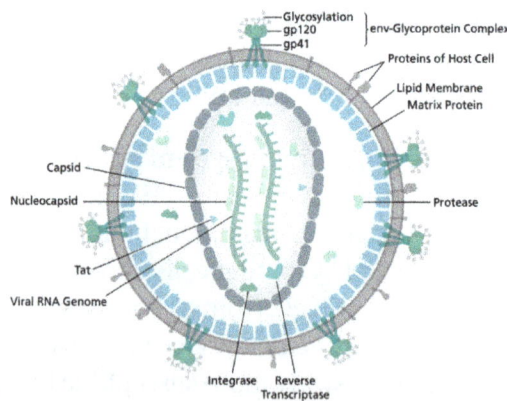

Diagram of HIV virion

HIV is different in structure from other retroviruses. It is roughly spherical with a diameter of about 120 nm, around 60 times smaller than a red blood cell. It is composed of two copies of positive single-stranded RNA that codes for the virus's nine genes enclosed by a conical capsid composed of 2,000 copies of the viral protein p24. The single-stranded RNA is tightly bound to nucleocapsid proteins, p7, and enzymes needed for the development of the virion such as reverse transcriptase, proteases, ribonuclease and integrase. A matrix composed of the viral protein p17 surrounds the capsid ensuring the integrity of the virion particle.

This is, in turn, surrounded by the viral envelope, that is composed of the lipid bilayer taken from the membrane of a human cell when the newly formed virus particle buds from the cell. The viral envelope contains proteins from the host cell and relatively few copies of the HIV Envelope protein, which consists of a cap made of three molecules known as glycoprotein (gp) 120, and a stem consisting of three gp41 molecules which anchor the structure into the viral envelope. The Envelope protein, encoded by the HIV *env* gene, allows the virus to attach to target cells and fuse the viral envelope with the target cell membrane releasing the viral contents into the cell and initiating the infectious cycle. As the sole viral protein on the surface of the virus, the Envelope protein is a major target for HIV vaccine efforts. Over half of the mass of the trimeric envelope spike is N-linked glycans. The density is high as the glycans shield the underlying viral protein from neutralisation by antibodies. This is one of the most densely glycosylated molecules known and the density is sufficiently high to prevent the normal maturation process of glycans during biogenesis in the endoplasmic and Golgi apparatus. The majority of the glycans are therefore stalled as immature 'high-mannose' glycans not normally present on secreted or cell surface human glycoproteins. The unusual processing and high density means that almost all broadly neutralising antibodies that have so far been identified (from a subset of patients that have been infected for many months to years) bind to or, are adapted to cope with, these envelope glycans.

The molecular structure of the viral spike has now been determined by X-ray crystallography and cryo-electron microscopy. These advances in structural biology were made possible due to the development of stable recombinant forms of the viral spike by the introduction of an intersubunit disulphide bond and an isoleucine to proline mutation in gp41. The so-called SOSIP trimers not only reproduce the antigenic properties of the native viral spike but also display the same degree of immature glycans as presented on the native virus. Recombinant trimeric viral spikes are promising vaccine candidates as they display less non-neutralising epitopes than recombinant monomeric gp120 which act to suppress the immune response to target epitopes.

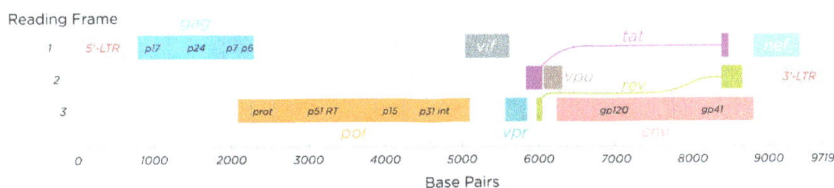

Structure of the RNA genome of HIV-1

The RNA genome consists of at least seven structural landmarks (LTR, TAR, RRE, PE, SLIP, CRS, and INS), and nine genes (*gag, pol,* and *env, tat, rev, nef, vif, vpr, vpu,* and sometimes a tenth *tev,* which is a fusion of tat env and rev), encoding 19 proteins. Three of these genes, *gag, pol,* and *env,* contain information needed to make the structural proteins for new virus particles. For example, *env* codes for a protein called gp160 that is cut in two by a cellular protease to form gp120 and gp41. The six remaining genes, *tat*, *rev*, *nef*, *vif*, *vpr*, and *vpu* (or *vpx* in the case of HIV-2), are regulatory genes for proteins that control the ability of HIV to infect cells, produce new copies of virus (replicate), or cause disease.

The two Tat proteins (p16 and p14) are transcriptional transactivators for the LTR promoter acting by binding the TAR RNA element. The TAR may also be processed into microRNAs that regulate the apoptosis genes ERCC1 and IER3. The Rev protein (p19) is involved in shuttling RNAs from the nucleus and the cytoplasm by binding to the RRE RNA element. The Vif protein (p23) prevents the action of APOBEC3G (a cellular protein that deaminates Cytidine to Uridine in the single stranded viral DNA and/or interferes with reverse transcription). The Vpr protein (p14) arrests cell division at G2/M. The Nef protein (p27) down-regulates CD4 (the major viral receptor), as well as the MHC class I and class II molecules.

Nef also interacts with SH3 domains. The Vpu protein (p16) influences the release of new virus particles from infected cells. The ends of each strand of HIV RNA contain an RNA sequence called the long terminal repeat (LTR). Regions in the LTR act as switches to control production of new viruses and can be triggered by proteins from either HIV or the host cell. The Psi element is involved in viral genome packaging and recognized by Gag and Rev proteins. The SLIP element (TTTTTT) is involved in the frameshift in the Gag-Pol reading frame required to make functional Pol.

Tropism

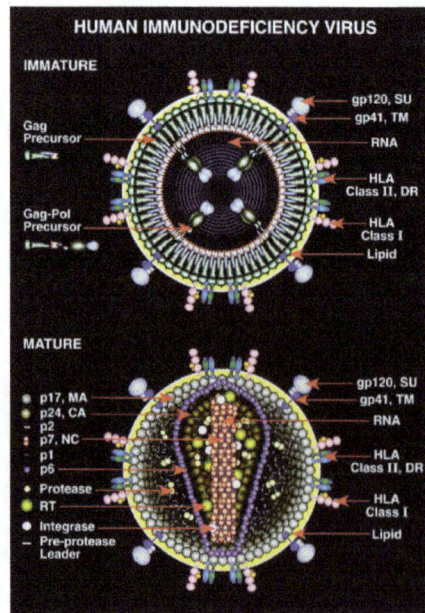

Diagram of the immature and mature forms of HIV

The term viral tropism refers to the cell types a virus infects. HIV can infect a variety of immune cells such as CD4⁺ T cells, macrophages, and microglial cells. HIV-1 entry to macrophages and CD4$^+$ T cells is mediated through interaction of the virion envelope glycoproteins (gp120) with the CD4 molecule on the target cells and also with chemokine coreceptors.

Macrophage (M-tropic) strains of HIV-1, or non-syncytia-inducing strains (NSI; now called R5 viruses) use the β-chemokine receptor CCR5 for entry and are, thus, able to replicate in macrophages and CD4$^+$ T cells. This CCR5 coreceptor is used by almost all primary HIV-1 isolates regardless of viral genetic subtype. Indeed, macrophages play a key role in several critical aspects of HIV infection. They appear to be the first cells infected by HIV and perhaps the source of HIV production when CD4$^+$ cells become depleted in the patient. Macrophages and microglial cells are the cells infected by HIV in the central nervous system. In tonsils and adenoids of HIV-infected patients, macrophages fuse into multinucleated giant cells that produce huge amounts of virus.

T-tropic isolates, or syncytia-inducing (SI; now called X4 viruses) strains replicate in primary CD4$^+$ T cells as well as in macrophages and use the α-chemokine receptor, CXCR4, for entry. Dual-tropic HIV-1 strains are thought to be transitional strains of HIV-1 and thus are able to use both CCR5 and CXCR4 as co-receptors for viral entry.

The α-chemokine SDF-1, a ligand for CXCR4, suppresses replication of T-tropic HIV-1 isolates. It does this by down-regulating the expression of CXCR4 on the surface of these cells. HIV that use only the CCR5 receptor are termed R5; those that use only CXCR4 are termed X4, and those that use both, X4R5. However, the use of coreceptor alone does not explain viral tropism, as not all R5 viruses are able to use CCR5 on macrophages for a productive infection and HIV can also infect a subtype of myeloid dendritic cells, which probably constitute a reservoir that maintains infection when CD4$^+$ T cell numbers have declined to extremely low levels.

Some people are resistant to certain strains of HIV. For example, people with the CCR5-Δ32 mutation are resistant to infection with R5 virus, as the mutation stops HIV from binding to this coreceptor, reducing its ability to infect target cells.

Sexual intercourse is the major mode of HIV transmission. Both X4 and R5 HIV are present in the seminal fluid, which is passed from a male to his sexual partner. The virions can then infect numerous cellular targets and disseminate into the whole organism. However, a selection process leads to a predominant transmission of the R5 virus through this pathway. How this selective process works is still under investigation, but one model is that spermatozoa may selectively carry R5 HIV as they possess both CCR3 and CCR5 but not CXCR4 on their surface and that genital epithelial cells preferentially sequester X4 virus. In patients infected with subtype B HIV-1, there is often a co-receptor switch in late-stage disease and T-tropic variants appear that can infect a variety of T cells through CXCR4. These variants then replicate more aggressively with heightened virulence that causes rapid T cell depletion, immune system collapse, and opportunistic infections that mark the advent of AIDS. Thus, during the course of infection, viral adaptation to the use of CXCR4 instead of CCR5 may be a key step in the progression to AIDS. A number of studies with subtype B-infected individuals have determined that between 40 and 50 percent of AIDS patients can harbour viruses of the SI and, it is presumed, the X4 phenotypes.

HIV-2 is much less pathogenic than HIV-1 and is restricted in its worldwide distribution. The adoption of "accessory genes" by HIV-2 and its more promiscuous pattern of coreceptor usage (including CD4-independence) may assist the virus in its adaptation to avoid innate restriction factors present in host cells. Adaptation to use normal cellular machinery to enable transmission and productive infection has also aided the establishment of HIV-2 replication in humans. A survival strategy for any infectious agent is not to kill its host but ultimately become a commensal organism. Having achieved a low pathogenicity, over time, variants more successful at transmission will be selected.

Replication Cycle

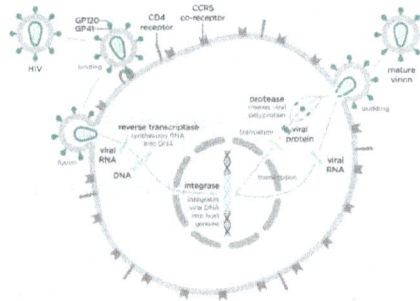

The HIV replication cycle

Entry to The Cell

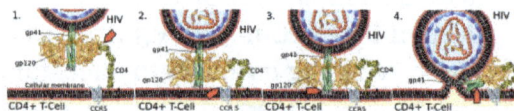

Mechanism of viral entry

1. Initial interaction between gp120 and CD4. 2. Conformational change in gp120 allows for secondary interaction with CCR5. 3. The distal tips of gp41 are inserted into the cellular membrane. 4. gp41 undergoes significant conformational change; folding in half and forming coiled-coils. This process pulls the viral and cellular membranes together, fusing them.

The HIV virion enters macrophages and CD4$^+$ T cells by the adsorption of glycoproteins on its surface to receptors on the target cell followed by fusion of the viral envelope with the cell membrane and the release of the HIV capsid into the cell.

Entry to the cell begins through interaction of the trimeric envelope complex (gp160 spike) and both CD4 and a chemokine receptor (generally either CCR5 or CXCR4, but others are known to interact) on the cell surface. gp120 binds to integrin $\alpha_4\beta_7$ activating LFA-1 the central integrin involved in the establishment of virological synapses, which facilitate efficient cell-to-cell spreading of HIV-1. The gp160 spike contains binding domains for both CD4 and chemokine receptors.

The first step in fusion involves the high-affinity attachment of the CD4 binding domains of gp120 to CD4. Once gp120 is bound with the CD4 protein, the envelope complex undergoes a structural change, exposing the chemokine binding domains of gp120 and allowing them to interact with the target chemokine receptor. This allows for a more stable two-pronged attachment, which allows the N-terminal fusion peptide gp41 to penetrate the cell membrane. Repeat sequences in gp41, HR1, and HR2 then interact, causing the collapse of the extracellular portion of gp41 into a hairpin. This loop structure brings the virus and cell membranes close together, allowing fusion of the membranes and subsequent entry of the viral capsid.

After HIV has bound to the target cell, the HIV RNA and various enzymes, including reverse transcriptase, integrase, ribonuclease, and protease, are injected into the cell. During the microtubule-based transport to the nucleus, the viral single-strand RNA genome is transcribed into double-strand DNA, which is then integrated into a host chromosome.

HIV can infect dendritic cells (DCs) by this CD4-CCR5 route, but another route using mannose-specific C-type lectin receptors such as DC-SIGN can also be used. DCs are one of the first cells encountered by the virus during sexual transmission. They are currently thought to play an important role by transmitting HIV to T-cells when the virus is captured in the mucosa by DCs. The presence of FEZ-1, which occurs naturally in neurons, is believed to prevent the infection of cells by HIV.

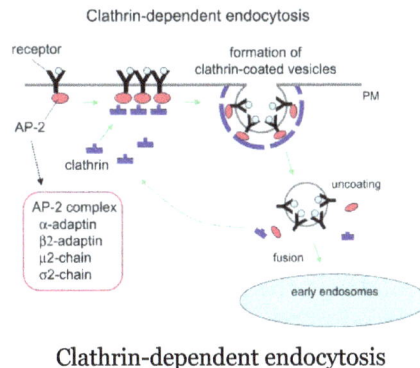

Clathrin-dependent endocytosis

HIV-1 entry, as well as entry of many other retroviruses, has long been believed to occur exclusively at the plasma membrane. More recently, however, productive infection by pH-independent, clathrin-dependent endocytosis of HIV-1 has also been reported and was recently suggested to constitute the only route of productive entry.

Replication and Transcription

Shortly after the viral capsid enters the cell, an enzyme called *reverse transcriptase* liberates the single-stranded (+)RNA genome from the attached viral proteins and copies it into a complementary DNA (cDNA) molecule. The process of reverse transcription is extremely error-prone, and the resulting mutations may cause drug resistance or allow the virus to evade the body's immune system. The reverse transcriptase also has ribonuclease activity that degrades the viral RNA during the synthesis of cDNA, as well as DNA-dependent DNA polymerase activity that creates a sense DNA from the *antisense* cDNA. Together, the cDNA and its complement form a double-stranded viral DNA that is then transported into the cell nucleus. The integration of the viral DNA into the host cell's genome is carried out by another viral enzyme called *integrase*.

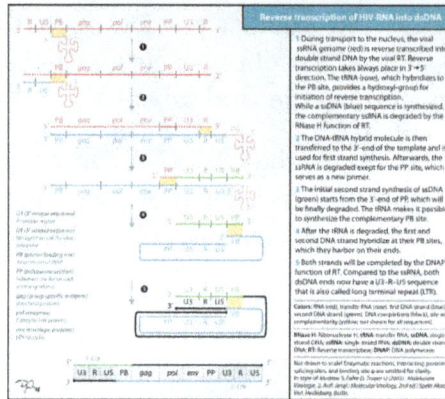

Reverse transcription of the HIV genome into double strand DNA

This integrated viral DNA may then lie dormant, in the latent stage of HIV infection. To actively produce the virus, certain cellular transcription factors need to be present, the most important of which is NF-κB (NF kappa B), which is upregulated when T-cells become activated. This means that those cells most likely to be killed by HIV are those currently fighting infection.

During viral replication, the integrated DNA provirus is transcribed into RNA, some of which then undergo RNA splicing to produce mature mRNAs. These mRNAs are exported from the nucleus into the cytoplasm, where they are translated into the regulatory proteins Tat (which encourages new virus production) and Rev. As the newly produced Rev protein accumulates in the nucleus, it binds to full-length, unspliced copies of virus RNAs and allows them to leave the nucleus. Some of these full-length RNAs function as new copies of the virus genome, while others function as mRNAs that are translated to produce the structural proteins Gag and Env. Gag proteins bind to copies of the virus RNA genome to package them into new virus particles.

HIV-1 and HIV-2 appear to package their RNA differently. HIV-1 will bind to any appropriate RNA. HIV-2 will preferentially bind to the mRNA that was used to create the Gag protein itself.

Recombination

Two RNA genomes are encapsidated in each HIV-1 particle (Structure and genome of HIV). Upon infection and replication catalyzed by reverse transcriptase, recombination between the two genomes can occur. Recombination occurs as the single-strand (+)RNA genomes are reverse transcribed to form DNA. During reverse transcription the nascent DNA can switch multiple times between the two copies of the viral RNA. This form of recombination is known as copy-choice. Recombination events may occur throughout the genome. From 2 to 20 events per genome may occur at each replication cycle, and these events can rapidly shuffle the genetic information that is transmitted from parental to progeny genomes.

Viral recombination produces genetic variation that likely contributes to the evolution of resistance to anti-retroviral therapy. Recombination may also contribute, in principle, to overcoming the immune defenses of the host. Yet, for the adaptive advantages of genetic variation to be realized, the two viral genomes packaged in individual infecting virus particles need to have arisen from separate progenitor parental viruses of differing genetic constitution. It is unknown how often such mixed packaging occurs under natural conditions.

Bonhoeffer et al. suggested that template switching by the reverse transcriptase acts as a repair process to deal with breaks in the ssRNA genome. In addition, Hu and Temin suggested that recombination is an adaptation for repair of damage in the RNA genomes. Strand switching (copy-choice recombination) by reverse transcriptase could generate an undamaged copy of genomic DNA from two damaged ssRNA genome copies. This view of the adaptive benefit of recombination in HIV could explain why each HIV particle contains two complete genomes, rather than one. Furthermore, the view that recombination is a repair process implies that the benefit of repair can occur at each replication cycle, and that this benefit can be realized whether or not the two genomes differ genetically. On the view that that recombination in HIV is a repair process, the generation of recombinational variation would be a consequence, but not the cause of, the evolution of template switching.

HIV-1 infection causes chronic ongoing inflammation and production of reactive oxygen species. Thus, the HIV genome may be vulnerable to oxidative damages, including breaks in the single-stranded RNA. For HIV, as well as for viruses generally, successful infection depends on overcoming host defensive strategies that often include production of genome-damaging reactive oxygen. Thus, Michod et al. suggested that recombination by viruses is an adaptation for repair of genome damages, and that recombinational variation is a byproduct that may provide a separate benefit.

Assembly and Release

The final step of the viral cycle, assembly of new HIV-1 virions, begins at the plasma membrane of the host cell. The Env polyprotein (gp160) goes through the endoplasmic reticulum and is transported to the Golgi complex where it is cleaved by furin resulting in the two HIV envelope glycoproteins, gp41 and gp120. These are transported to the plasma membrane of the host cell where gp41 anchors gp120 to the membrane of the infected cell. The Gag (p55) and Gag-Pol (p160) polyproteins also associate with the inner surface of the plasma membrane along with the HIV genomic RNA as the forming virion begins to bud from the host cell. The budded virion is still immature as the gag polyproteins still need to be cleaved into the actual matrix, capsid and nucleocapsid proteins. This cleavage is mediated by the also packaged viral protease and can be inhibited by antiretroviral drugs of the protease inhibitor class. The various structural components then assemble to produce a mature HIV virion. Only mature virions are then able to infect another cell.

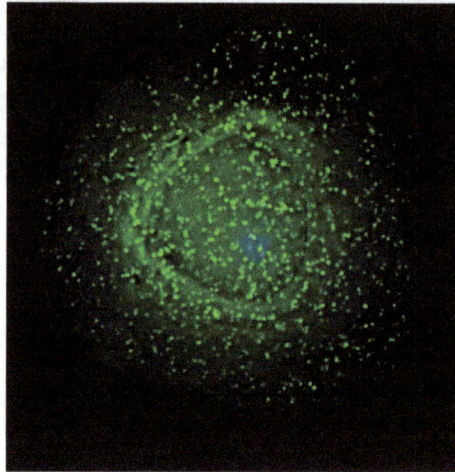

HIV assembling on the surface of an infected macrophage.

Spread Within The Body

HIV is now known to spread between CD4+ T cells by two parallel routes: cell-free spread and cell-to-cell spread, i.e. it employs hybrid spreading mechanisms. In the cell-free spread, virus particles bud from an infected T cell, enter the blood/extracellular fluid and then infect another T cell following a chance encounter. HIV can also disseminate by direct transmission from one cell to another by a process of cell-to-cell spread. Two pathways of cell-to-cell transmission have been reported. Firstly, an infected T cell can transmit virus directly to a target T cell via a virological synapse. Secondly, an antigen presenting cell (APC) can also transmit HIV to T cells by a process that either involves productive infection (in the case of macrophages) or capture and transfer of virions *in trans* (in the case of dendritic cells). Whichever pathway is used, infection by cell-to-cell transfer is reported to be much more efficient than cell-free virus spread. A number of factors contribute to this increased efficiency, including polarised virus budding towards the site of cell-to-cell contact, close apposition of cells which minimizes fluid-phase diffusion of virions, and clustering of HIV entry receptors on the target cell to the contact zone. Cell-to-cell spread is thought to be particularly important in lymphoid tissues where CD4+ T lymphocytes are densely packed and likely to frequently interact. Intravital imaging studies have supported the concept of the HIV virological synapse *in vivo*. The hybrid spreading mechanisms of HIV contribute to the virus's ongoing replication against antiretroviral therapies.

Genetic Variability

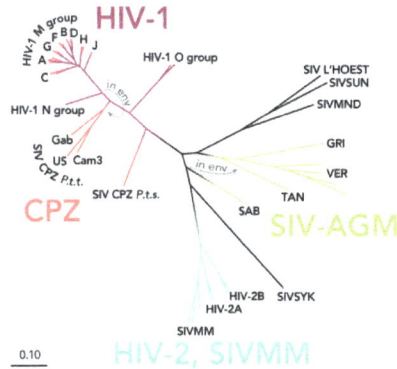

The phylogenetic tree of the SIV and HIV

HIV differs from many viruses in that it has very high genetic variability. This diversity is a result of its fast replication cycle, with the generation of about 10^{10} virions every day, coupled with a high mutation rate of approximately 3×10^{-5} per nucleotide base per cycle of replication and recombinogenic properties of reverse transcriptase.

This complex scenario leads to the generation of many variants of HIV in a single infected patient in the course of one day. This variability is compounded when a single cell is simultaneously infected by two or more different strains of HIV. When simultaneous infection occurs, the genome of progeny virions may be composed of RNA strands from two different strains. This hybrid virion then infects a new cell where it undergoes replication. As this happens, the reverse transcriptase, by jumping back and forth between the two different RNA templates, will generate a newly synthesized retroviral DNA sequence that is a recombinant between the two parental genomes. This recombination is most obvious when it occurs between subtypes.

The closely related simian immunodeficiency virus (SIV) has evolved into many strains, classified by the natural host species. SIV strains of the African green monkey (SIVagm) and sooty mangabey (SIVsmm) are thought to have a long evolutionary history with their hosts. These hosts have adapted to the presence of the virus, which is present at high levels in the host's blood but evokes only a mild immune response, does not cause the development of simian AIDS, and does not undergo the extensive mutation and recombination typical of HIV infection in humans.

In contrast, when these strains infect species that have not adapted to SIV ("heterologous" hosts such as rhesus or cynomologus macaques), the animals develop AIDS and the virus generates genetic diversity similar to what is seen in human HIV infection. Chimpanzee SIV (SIVcpz), the closest genetic relative of HIV-1, is associated with increased mortality and AIDS-like symptoms in its natural host. SIVcpz appears to have been transmitted relatively recently to chimpanzee and human populations, so their hosts have not yet adapted to the virus. This virus has also lost a function of the Nef gene that is present in most SIVs. For non-pathogenic SIV variants, Nef suppresses T-cell activation through the CD3 marker. Nef's function in non-pathogenic forms of SIV is to downregulate expression of inflammatory cytokines, MHC-1, and signals that affect T cell trafficking. In HIV-1 and SIVcpz, Nef does not inhibit T-cell activation and it has lost this function. Without this function, T cell depletion is more likely, leading to immunodeficiency.

Three groups of HIV-1 have been identified on the basis of differences in the envelope (*env*) region: M, N, and O. Group M is the most prevalent and is subdivided into eight subtypes (or clades), based on the whole genome, which are geographically distinct. The most prevalent are subtypes B (found mainly in North America and Europe), A and D (found mainly in Africa), and C (found mainly in Africa and Asia); these subtypes form branches in the phylogenetic tree representing the lineage of the M group of HIV-1. Coinfection with distinct subtypes gives rise to circulating recombinant forms (CRFs). In 2000, the last year in which an analysis of global subtype prevalence was made, 47.2% of infections worldwide were of subtype C, 26.7% were of subtype A/CRF02_AG, 12.3% were of subtype B, 5.3% were of subtype D, 3.2% were of CRF_AE, and the remaining 5.3% were composed of other subtypes and CRFs. Most HIV-1 research is focused on subtype B; few laboratories focus on the other subtypes. The existence of a fourth group, "P", has been hypothesised based on a virus isolated in 2009. The strain is apparently derived from gorilla SIV (SIVgor), first isolated from western lowland gorillas in 2006.

HIV-2's closest relative is SIVsm, a strain of SIV found in sooty mangabees. Since HIV-1 is derived from SIVcpz, and HIV-2 from SIVsm, the genetic sequence of HIV-2 is only partially homologous to HIV-1 and more closely resembles that of SIVsm.

Diagnosis

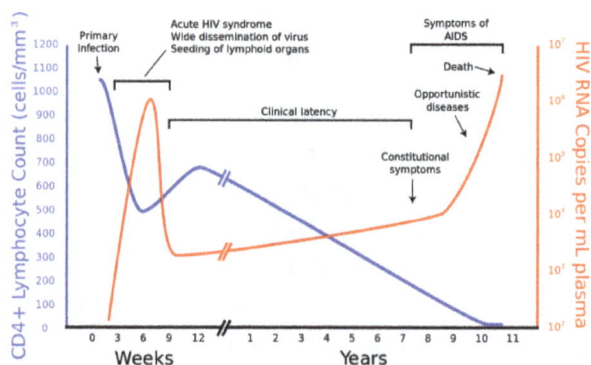

A generalized graph of the relationship between HIV copies (viral load) and CD4 counts over the average course of untreated HIV infection; any particular individual's disease course may vary considerably.

CD4⁺ T cell count (cells per μL)

HIV RNA copies per mL of plasma

Many HIV-positive people are unaware that they are infected with the virus. For example, in 2001 less than 1% of the sexually active urban population in Africa had been tested, and this proportion is even lower in rural populations. Furthermore, in 2001 only 0.5% of pregnant women attending urban health facilities were counselled, tested or receive their test results. Again, this proportion is even lower in rural health facilities. Since donors may therefore be unaware of their infection, donor blood and blood products used in medicine and medical research are routinely screened for HIV.

HIV-1 testing is initially by an enzyme-linked immunosorbent assay (ELISA) to detect antibodies to HIV-1. Specimens with a nonreactive result from the initial ELISA are considered HIV-negative unless new exposure to an infected partner or partner of unknown HIV status has occurred. Specimens with a reactive ELISA result are retested in duplicate. If the result of either duplicate test is reactive, the specimen is reported as repeatedly reactive and undergoes confirmatory testing with a more specific supplemental test (e.g., western blot or, less commonly, an immunofluorescence assay (IFA)). Only specimens that are repeatedly reactive by ELISA and positive by IFA or reactive by western blot are considered HIV-positive and indicative of HIV infection. Specimens that are repeatedly ELISA-reactive occasionally provide an indeterminate western blot result, which may be either an incomplete antibody response to HIV in an infected person or nonspecific reactions in an uninfected person.

Although IFA can be used to confirm infection in these ambiguous cases, this assay is not widely used. In general, a second specimen should be collected more than a month later and retested for persons with indeterminate western blot results. Although much less commonly available, nucleic acid testing (e.g., viral RNA or proviral DNA amplification method) can also help diagnosis in certain situations. In addition, a few tested specimens might provide inconclusive results because of a low quantity specimen. In these situations, a second specimen is collected and tested for HIV infection.

Modern HIV testing is extremely accurate. A single screening test is correct more than 99% of the time. The chance of a false-positive result in standard two-step testing protocol is estimated to be about 1 in 250,000 in a low risk population. Testing post exposure is recommended initially and at six weeks, three months, and six months.

The latest recommendations of the CDC show that HIV testing must start with an immunoassay combination test for HIV-1 and HIV-2 antibodies and p24 antigen. A negative result rules out HIV exposure, while a positive one must be followed by an HIV-1/2 antibody differentiation immunoassay to detect which is present. This gives rise to four possible scenarios:

- 1. HIV-1 (+) & HIV-2 (-): HIV-1 antibodies detected

- 2. HIV-1 (-) & HIV-2 (+): HIV-2 antibodies detected

- 3. HIV-1 (+) & HIV-2 (+): HIV antibodies detected

- 4. HIV-1 (-) or indeterminate & HIV-2 (-): Nucleic acid test must be carried out to detect the acute infection of HIV-1 or its absence.

An updated algorithm published by the CDC in June 2014 recommends that diagnosis starts with the p24 antigen test. A negative result rules out infection, while a positive one must be followed by an HIV-1/2 antibody differentiation immunoassay. A positive differentiation test confirms diagnosis, while a negative or indeterminate result must be followed by nucleic acid test (NAT). A positive NAT result confirms HIV-1 infection whereas a negative result rules out infection (false positive p24).

Research

HIV/AIDS research includes all medical research that attempts to prevent, treat, or cure HIV/AIDS, as well as fundamental research about the nature of HIV as an infectious agent and AIDS as the disease caused by HIV.

Many governments and research institutions participate in HIV/AIDS research. This research includes behavioral health interventions, such as research into sex education, and drug development, such as research into microbicides for sexually transmitted diseases, HIV vaccines, and antiretroviral drugs. Other medical research areas include the topics of pre-exposure prophylaxis, post-exposure prophylaxis, circumcision and HIV, and accelerated aging effects.

History

Discovery

AIDS was first clinically observed in 1981 in the United States. The initial cases were a cluster of injection drug users and gay men with no known cause of impaired immunity who showed symptoms of *Pneumocystis carinii* pneumonia (PCP), a rare opportunistic infection that was known to occur in people with very compromised immune systems. Soon thereafter, additional gay men developed a previously rare skin cancer called Kaposi's sarcoma (KS). Many more cases of PCP and KS emerged, alerting U.S. Centers for Disease Control and Prevention (CDC) and a CDC task force was formed to monitor the outbreak. The earliest retrospectively described case of AIDS is believed to have been in Norway beginning in 1966.

In the beginning, the CDC did not have an official name for the disease, often referring to it by way of the diseases that were associated with it, for example, lymphadenopathy, the disease after which the discoverers of HIV originally named the virus. They also used *Kaposi's Sarcoma and Opportunistic Infections*, the name by which a task force had been set up in 1981. In the general press, the term *GRID*, which stood for gay-related immune deficiency, had been coined. The CDC, in search of a name, and looking at the infected communities coined "the 4H disease," as it seemed to single out homosexuals, heroin users, hemophiliacs, and Haitians. However, after determining that AIDS was not isolated to the gay community, it was realized that the term GRID was misleading and *AIDS* was introduced at a meeting in July 1982. By September 1982 the CDC started using the name AIDS.

Robert Gallo, co-discoverer of HIV

In 1983, two separate research groups led by Robert Gallo and Luc Montagnier independently declared that a novel retrovirus may have been infecting AIDS patients, and published their findings in the same issue of the journal *Science*. Gallo claimed that a virus his group had isolated from a person with AIDS was strikingly similar in shape to other human T-lymphotropic viruses (HTLVs) his group had been the first to isolate. Gallo's group called their newly isolated virus HTLV-III. At the same time, Montagnier's group isolated a virus from a patient presenting with swelling of the lymph nodes of the neck and physical weakness, two classic symptoms of AIDS. Contradicting the report from Gallo's group, Montagnier and his colleagues showed that core proteins of this virus were immunologically different from those of HTLV-I. Montagnier's group named their isolated virus lymphadenopathy-associated virus (LAV). As these two viruses turned out to be the same, in 1986, LAV and HTLV-III were renamed HIV.

Origins

Both HIV-1 and HIV-2 are believed to have originated in non-human primates in West-central Africa, and are believed to have transferred to humans (a process known as zoonosis) in the early 20th century.

HIV-1 appears to have originated in southern Cameroon through the evolution of SIV(cpz), a simian immunodeficiency virus (SIV) that infects wild chimpanzees (HIV-1 descends from the SIV(cpz) endemic in the chimpanzee subspecies *Pan troglodytes troglodytes*). The closest relative of HIV-2 is SIV (smm), a virus of the sooty mangabey (*Cercocebus atys atys*), an Old World monkey living in litoral West Africa (from southern Senegal to western Côte d'Ivoire). New World monkeys such as the owl monkey are resistant to HIV-1 infection, possibly because of a genomic fusion of two viral resistance genes. HIV-1 is thought to have jumped the species barrier on at least three separate occasions, giving rise to the three groups of the virus, M, N, and O.

Left to right: the African green monkey source of SIV, the sooty mangabey source of HIV-2, and the chimpanzee source of HIV-1

There is evidence that humans who participate in bushmeat activities, either as hunters or as bushmeat vendors, commonly acquire SIV. However, SIV is a weak virus, and it is typically suppressed by the human immune system within weeks of infection. It is thought that several transmissions of the virus from individual to individual in quick succession are necessary to allow it enough time to mutate into HIV. Furthermore, due to its relatively low person-to-person transmission rate, it can only spread throughout the population in the presence of one or more of high-risk transmission channels, which are thought to have been absent in Africa prior to the 20th century.

Specific proposed high-risk transmission channels, allowing the virus to adapt to humans and spread throughout the society, depend on the proposed timing of the animal-to-human crossing. Genetic studies of the virus suggest that the most recent common ancestor of the HIV-1 M group dates back to circa 1910. Proponents of this dating link the HIV epidemic with the emergence of colonialism and growth of large colonial African cities, leading to social changes, including a higher degree of sexual promiscuity, the spread of prostitution, and the concomitant high frequency of genital ulcer diseases (such as syphilis) in nascent colonial cities. While transmission rates of HIV during vaginal intercourse are typically low, they are increased many fold if one of the partners suffers from a sexually transmitted infection resulting in genital ulcers. Early 1900s colonial cities were notable due to their high prevalence of prostitution and genital ulcers to the degree that as of 1928 as many as 45% of female residents of eastern Leopoldville were thought to have been prostitutes and as of 1933 around 15% of all residents of the same city were infected by one of the forms of syphilis.

An alternative view holds that unsafe medical practices in Africa during years following World War II, such as unsterile reuse of single use syringes during mass vaccination, antibiotic, and anti-malaria treatment campaigns, were the initial vector that allowed the virus to adapt to humans and spread.

The earliest well documented case of HIV in a human dates back to 1959 in the Belgian Congo. The virus may have been present in the United States as early as the mid-to-late 1950s, as a sixteen-year-old male presented with symptoms in 1966 died in 1969.

HIV/AIDS

Human immunodeficiency virus infection and acquired immune deficiency syndrome (HIV/AIDS) is a spectrum of conditions caused by infection with the human immunodeficiency virus (HIV). Following initial infection, a person may not notice any symptoms or may experience a brief period of influenza-like illness. Typically, this is followed by a prolonged period with no symptoms. As the infection progresses, it interferes more with the immune system, increasing the risk of common infections like tuberculosis, as well as other opportunistic infections, and tumors that rarely affect people who have working immune systems. These late symptoms of infection are referred to as AIDS. This stage is often also associated with weight loss.

HIV is spread primarily by unprotected sex (including anal and oral sex), contaminated blood transfusions, hypodermic needles, and from mother to child during pregnancy, delivery, or breastfeeding. Some bodily fluids, such as saliva and tears, do not transmit HIV. Methods of prevention include safe sex, needle exchange programmes, treating those who are infected, and male circumcision. Disease in a baby can often be prevented by giving both the mother and child antiretroviral medication. There is no cure or vaccine; however, antiretroviral treatment can slow the course of the disease and may lead to a near-normal life expectancy. Treatment is recommended as soon as the diagnosis is made. Without treatment, the average survival time after infection is 11 years.

In 2014 about 36.9 million people were living with HIV and it resulted in 1.2 million deaths. Most of those infected live in sub-Saharan Africa. Between its discovery and 2014 AIDS has caused an estimated 39 million deaths worldwide. HIV/AIDS is considered a pandemic—a disease outbreak which is present over a large area and is actively spreading. HIV is believed to have originated in west-central Africa during the late 19th or early 20th century. AIDS was first recognized by the United States Centers for Disease Control and Prevention (CDC) in 1981 and its cause—HIV infection—was identified in the early part of the decade.

HIV/AIDS has had a great impact on society, both as an illness and as a source of discrimination. The disease also has large economic impacts. There are many misconceptions about HIV/AIDS such as the belief that it can be transmitted by casual non-sexual contact. The disease has become subject to many controversies involving religion including the Catholic church's decision not to support condom use as prevention. It has attracted international medical and political attention as well as large-scale funding since it was identified in the 1980s.

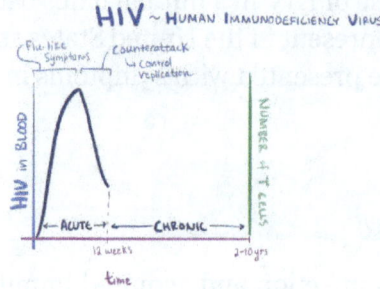

HIV ~ HUMAN IMMUNODEFICIENCY VIRUS

Video explanation

Signs and Symptoms

There are three main stages of HIV infection: acute infection, clinical latency and AIDS.

Acute Infection

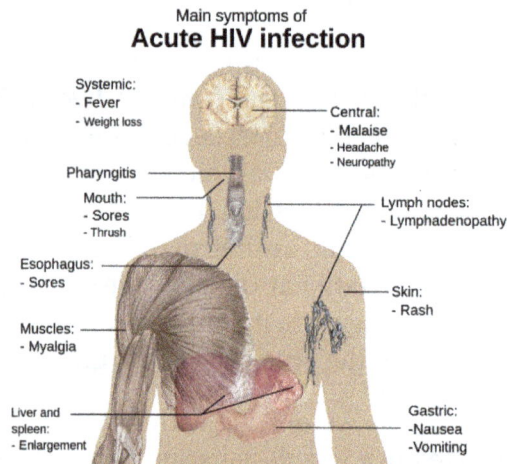

Main symptoms of
Acute HIV infection

Main symptoms of acute HIV infection

The initial period following the contraction of HIV is called acute HIV, primary HIV or acute retroviral syndrome. Many individuals develop an influenza-like illness or a mononucleosis-like illness 2–4 weeks post exposure while others have no significant symptoms. Symptoms occur in 40–90% of cases and most commonly include fever, large tender lymph nodes, throat inflammation, a rash, headache, and/or sores of the mouth and genitals. The rash, which occurs in 20–50% of cases, presents itself on the trunk and is maculopapular, classically. Some people also develop opportunistic infections at this stage. Gastrointestinal symptoms such as nausea, vomiting or diarrhea may occur, as may neurological symptoms of peripheral neuropathy or Guillain-Barre syndrome. The duration of the symptoms varies, but is usually one or two weeks.

Due to their nonspecific character, these symptoms are not often recognized as signs of HIV infection. Even cases that do get seen by a family doctor or a hospital are often misdiagnosed as one of the many common infectious diseases with overlapping symptoms. Thus, it is recommended that HIV be considered in people presenting an unexplained fever who may have risk factors for the infection.

Clinical Latency

The initial symptoms are followed by a stage called clinical latency, asymptomatic HIV, or chronic HIV. Without treatment, this second stage of the natural history of HIV infection can last from about three years to over 20 years (on average, about eight years). While typically there are few or no symptoms at first, near the end of this stage many people experience fever, weight loss, gastrointestinal problems and muscle pains. Between 50 and 70% of people also develop persistent generalized lymphadenopathy, characterized by unexplained, non-painful enlargement of more than one group of lymph nodes (other than in the groin) for over three to six months.

Although most HIV-1 infected individuals have a detectable viral load and in the absence of treatment will eventually progress to AIDS, a small proportion (about 5%) retain high levels of CD4$^+$ T cells (T helper cells) without antiretroviral therapy for more than 5 years. These individuals are classified as HIV controllers or long-term nonprogressors (LTNP). Another group consists of those who maintain a low or undetectable viral load without anti-retroviral treatment, known as "elite controllers" or "elite suppressors". They represent approximately 1 in 300 infected persons.

Acquired Immuno Deficiency Syndrome

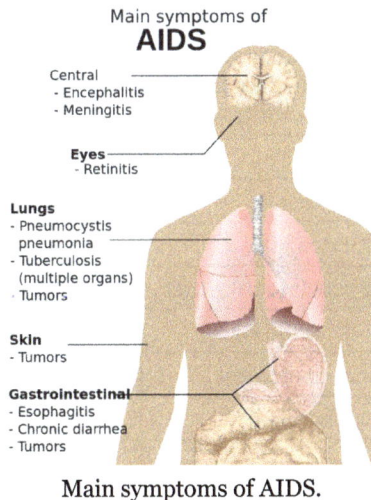

Main symptoms of
AIDS

Central
- Encephalitis
- Meningitis

Eyes
- Retinitis

Lungs
- Pneumocystis pneumonia
- Tuberculosis (multiple organs)
- Tumors

Skin
- Tumors

Gastrointestinal
- Esophagitis
- Chronic diarrhea
- Tumors

Main symptoms of AIDS.

Acquired immunodeficiency syndrome (AIDS) is defined in terms of either a CD4$^+$ T cell count below 200 cells per µL or the occurrence of specific diseases in association with an HIV infection. In the absence of specific treatment, around half of people infected with HIV develop AIDS within ten years. The most common initial conditions that alert to the presence of AIDS are pneumocystis pneumonia (40%), cachexia in the form of HIV wasting syndrome (20%), and esophageal candidiasis. Other common signs include recurring respiratory tract infections.

Opportunistic infections may be caused by bacteria, viruses, fungi, and parasites that are normally controlled by the immune system. Which infections occur depends partly on what organisms are common in the person's environment. These infections may affect nearly every organ system.

People with AIDS have an increased risk of developing various viral-induced cancers, including Kaposi's sarcoma, Burkitt's lymphoma, primary central nervous system lymphoma, and cervical cancer. Kaposi's sarcoma is the most common cancer occurring in 10 to 20% of people with HIV. The second most common cancer is lymphoma, which is the cause of death of nearly 16% of people with AIDS and is the initial sign of AIDS in 3 to 4%. Both these cancers are associated with human herpesvirus 8. Cervical cancer occurs more frequently in those with AIDS because of its association with human papillomavirus (HPV). Conjunctival cancer (of the layer that lines the inner part of eyelids and the white part of the eye) is also more common in those with HIV.

Additionally, people with AIDS frequently have systemic symptoms such as prolonged fevers, sweats (particularly at night), swollen lymph nodes, chills, weakness, and unintended weight loss. Diarrhea is another common symptom, present in about 90% of people with AIDS. They can also be affected by diverse psychiatric and neurological symptoms independent of opportunistic infections and cancers.

Transmission

Average per act risk of getting HIV by exposure route to an infected source	
Exposure route	**Chance of infection**
Blood transfusion	90%
Childbirth (to child)	25%
Needle-sharing injection drug use	0.67%
Percutaneous needle stick	0.30%
Receptive anal intercourse*	0.04–3.0%
Insertive anal intercourse*	0.03%
Receptive penile-vaginal intercourse*	0.05–0.30%
Insertive penile-vaginal intercourse*	0.01–0.38%
Receptive oral intercourse*§	0–0.04%
Insertive oral intercourse*§	0–0.005%
* assuming no condom use § source refers to oral intercourse performed on a man	

HIV is transmitted by three main routes: sexual contact, significant exposure to infected body fluids or tissues, and from mother to child during pregnancy, delivery, or breastfeeding (known as vertical transmission). There is no risk of acquiring HIV if exposed to feces, nasal secretions, saliva, sputum, sweat, tears, urine, or vomit unless these are contaminated with blood. It is possible to be co-infected by more than one strain of HIV—a condition known as HIV superinfection.

Sexual

The most frequent mode of transmission of HIV is through sexual contact with an infected person. The majority of all transmissions worldwide occur through heterosexual contacts (i.e. sexual contacts between people of the opposite sex); however, the pattern of transmission varies significantly among countries. In the United States, as of 2010, most transmission occurred in men who had sex with men, with this population accounting for 65% of all new cases.

With regard to unprotected heterosexual contacts, estimates of the risk of HIV transmission per sexual act appear to be four to ten times higher in low-income countries than in high-income countries. In low-income countries, the risk of female-to-male transmission is estimated as 0.38% per act, and of male-to-female transmission as 0.30% per act; the equivalent estimates for high-income countries are 0.04% per act for female-to-male transmission, and 0.08% per act for male-to-female transmission. The risk of transmission from anal intercourse is especially high, estimated as 1.4–1.7% per act in both heterosexual and homosexual contacts. While the risk of transmission from oral sex is relatively low, it is still present. The risk from receiving oral sex has been described as "nearly nil"; however, a few cases have been reported. The per-act risk is estimated at 0–0.04% for receptive oral intercourse. In settings involving prostitution in low income countries, risk of female-to-male transmission has been estimated as 2.4% per act and male-to-female transmission as 0.05% per act.

Risk of transmission increases in the presence of many sexually transmitted infections and genital ulcers. Genital ulcers appear to increase the risk approximately fivefold. Other sexually transmitted infections, such as gonorrhea, chlamydia, trichomoniasis, and bacterial vaginosis, are associated with somewhat smaller increases in risk of transmission.

The viral load of an infected person is an important risk factor in both sexual and mother-to-child transmission. During the first 2.5 months of an HIV infection a person's infectiousness is twelve times higher due to this high viral load. If the person is in the late stages of infection, rates of transmission are approximately eightfold greater.

Commercial sex workers (including those in pornography) have an increased rate of HIV. Rough sex can be a factor associated with an increased risk of transmission. Sexual assault is also believed to carry an increased risk of HIV transmission as condoms are rarely worn, physical trauma to the vagina or rectum is likely, and there may be a greater risk of concurrent sexually transmitted infections.

Body Fluids

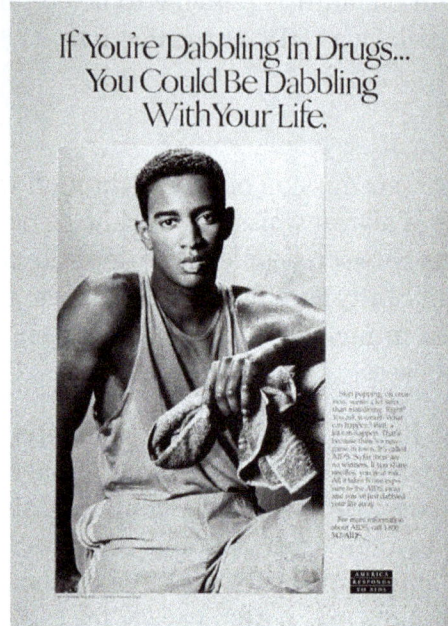

CDC poster from 1989 highlighting the threat of AIDS associated with drug use

The second most frequent mode of HIV transmission is via blood and blood products. Blood-borne transmission can be through needle-sharing during intravenous drug use, needle stick injury, transfusion of contaminated blood or blood product, or medical injections with unsterilised equipment. The risk from sharing a needle during drug injection is between 0.63 and 2.4% per act, with an average of 0.8%. The risk of acquiring HIV from a needle stick from an HIV-infected person is estimated as 0.3% (about 1 in 333) per act and the risk following mucous membrane exposure to infected blood as 0.09% (about 1 in 1000) per act. In the United States intravenous drug users made up 12% of all new cases of HIV in 2009, and in some areas more than 80% of people who inject drugs are HIV positive.

HIV is transmitted in about 93% of blood transfusions using infected blood. In developed countries the risk of acquiring HIV from a blood transfusion is extremely low (less than one in half a million) where improved donor selection and HIV screening is performed; for example, in the UK the risk is reported at one in five million and in the United States it was one in 1.5 million in 2008. In low income countries, only half of transfusions may be appropriately screened (as of 2008), and it is estimated that up to 15% of HIV infections in these areas come from transfusion of infected blood and blood products, representing between 5% and 10% of global infections. Although rare because of screening, it is possible to acquire HIV from organ and tissue transplantation.

Unsafe medical injections play a significant role in HIV spread in sub-Saharan Africa. In 2007, between 12 and 17% of infections in this region were attributed to medical syringe use. The World Health Organization estimates the risk of transmission as a result of a medical injection in Africa at 1.2%. Significant risks are also associated with invasive procedures, assisted delivery, and dental care in this area of the world.

People giving or receiving tattoos, piercings, and scarification are theoretically at risk of infection but no confirmed cases have been documented. It is not possible for mosquitoes or other insects to transmit HIV.

Mother-to-child

HIV can be transmitted from mother to child during pregnancy, during delivery, or through breast milk resulting in infection in the baby. This is the third most common way in which HIV is transmitted globally. In the absence of treatment, the risk of transmission before or during birth is around 20% and in those who also breast-feed 35%. As of 2008, vertical transmission accounted for about 90% of cases of HIV in children. With appropriate treatment the risk of mother-to-child infection can be reduced to about 1%. Preventive treatment involves the mother taking antiretrovirals during pregnancy and delivery, an elective caesarean section, avoiding breastfeeding, and administering antiretroviral drugs to the newborn. Antiretrovirals when taken by either the mother or the infant decrease the risk of transmission in those who do breastfeed. Many of these measures are however not available in the developing world. If blood contaminates food during pre-chewing it may pose a risk of transmission.

Virology

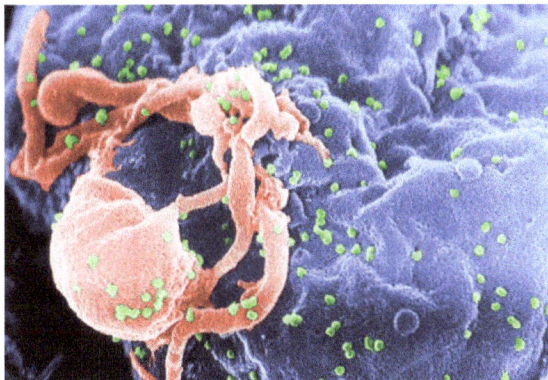

Scanning electron micrograph of HIV-1, colored green, budding from a cultured lymphocyte.

HIV is the cause of the spectrum of disease known as HIV/AIDS. HIV is a retrovirus that primarily infects components of the human immune system such as CD4+ T cells, macrophages and dendritic cells. It directly and indirectly destroys CD4+ T cells.

HIV is a member of the genus *Lentivirus*, part of the family *Retroviridae*. Lentiviruses share many morphological and biological characteristics. Many species of mammals are infected by lentiviruses, which are characteristically responsible for long-duration illnesses with a long incubation period. Lentiviruses are transmitted as single-stranded, positive-sense, enveloped RNA viruses. Upon entry into the target cell, the viral RNA genome is converted (reverse transcribed) into double-stranded DNA by a virally encoded reverse transcriptase that is transported along with the viral genome in the virus particle. The resulting viral DNA is then imported into the cell nucleus and integrated into the cellular DNA by a virally encoded integrase and host co-factors. Once integrated, the virus may become latent, allowing the virus and its host cell to avoid detection by the immune system. Alternatively, the virus may be transcribed, producing new RNA genomes and viral proteins that are packaged and released from the cell as new virus particles that begin the replication cycle anew.

HIV is now known to spread between CD4+ T cells by two parallel routes: cell-free spread and cell-to-cell spread, i.e. it employs hybrid spreading mechanisms. In the cell-free spread, virus particles bud from an infected T cell, enter the blood/extracellular fluid and then infect another T cell following a chance encounter. HIV can also disseminate by direct transmission from one cell to another by a process of cell-to-cell spread. The hybrid spreading mechanisms of HIV contribute to the virus's ongoing replication against antiretroviral therapies.

Two types of HIV have been characterized: HIV-1 and HIV-2. HIV-1 is the virus that was originally discovered (and initially referred to also as LAV or HTLV-III). It is more virulent, more infective, and is the cause of the majority of HIV infections globally. The lower infectivity of HIV-2 as compared with HIV-1 implies that fewer people exposed to HIV-2 will be infected per exposure. Because of its relatively poor capacity for transmission, HIV-2 is largely confined to West Africa.

Pathophysiology

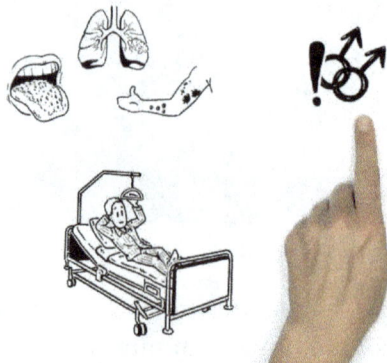

HIV/AIDS explained in a simple way

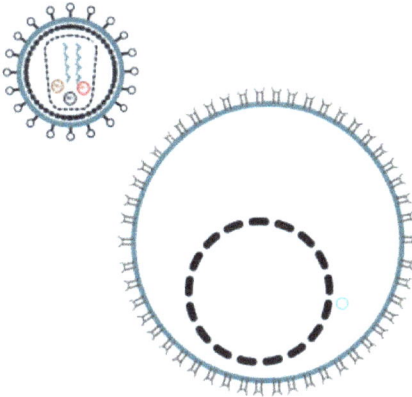

HIV replication cycle

After the virus enters the body there is a period of rapid viral replication, leading to an abundance of virus in the peripheral blood. During primary infection, the level of HIV may reach several million virus particles per milliliter of blood. This response is accompanied by a marked drop in the number of circulating $CD4^+$ T cells. The acute viremia is almost invariably associated with activation of $CD8^+$ T cells, which kill HIV-infected cells, and subsequently with antibody production, or seroconversion. The $CD8^+$ T cell response is thought to be important in controlling virus levels, which peak and then decline, as the $CD4^+$ T cell counts recover. A good $CD8^+$ T cell response has been linked to slower disease progression and a better prognosis, though it does not eliminate the virus.

Ultimately, HIV causes AIDS by depleting $CD4^+$ T cells. This weakens the immune system and allows opportunistic infections. T cells are essential to the immune response and without them, the body cannot fight infections or kill cancerous cells. The mechanism of $CD4^+$ T cell depletion differs in the acute and chronic phases. During the acute phase, HIV-induced cell lysis and killing of infected cells by cytotoxic T cells accounts for $CD4^+$ T cell depletion, although apoptosis may also be a factor. During the chronic phase, the consequences of generalized immune activation coupled with the gradual loss of the ability of the immune system to generate new T cells appear to account for the slow decline in $CD4^+$ T cell numbers.

Although the symptoms of immune deficiency characteristic of AIDS do not appear for years after a person is infected, the bulk of $CD4^+$ T cell loss occurs during the first weeks of infection, especially in the intestinal mucosa, which harbors the majority of the lymphocytes found in the body. The reason for the preferential loss of mucosal $CD4^+$ T cells is that the majority of mucosal $CD4^+$ T cells express the CCR5 protein which HIV uses as a co-receptor to gain access to the cells, whereas only a small fraction of $CD4^+$ T cells in the bloodstream do so. A specific genetic change that alters the CCR5 protein when present in both chromosomes very effectively prevents HIV-1 infection.

HIV seeks out and destroys CCR5 expressing CD4$^+$ T cells during acute infection. A vigorous immune response eventually controls the infection and initiates the clinically latent phase. CD4$^+$ T cells in mucosal tissues remain particularly affected. Continuous HIV replication causes a state of generalized immune activation persisting throughout the chronic phase. Immune activation, which is reflected by the increased activation state of immune cells and release of pro-inflammatory cytokines, results from the activity of several HIV gene products and the immune response to ongoing HIV replication. It is also linked to the breakdown of the immune surveillance system of the gastrointestinal mucosal barrier caused by the depletion of mucosal CD4$^+$ T cells during the acute phase of disease.

Diagnosis

HIV/AIDS is diagnosed via laboratory testing and then staged based on the presence of certain signs or symptoms. HIV screening is recommended by the United States Preventive Services Task Force for all people 15 years to 65 years of age including all pregnant women. Additionally, testing is recommended for those at high risk, which includes anyone diagnosed with a sexually transmitted illness. In many areas of the world, a third of HIV carriers only discover they are infected at an advanced stage of the disease when AIDS or severe immunodeficiency has become apparent.

HIV Testing

Most people infected with HIV develop specific antibodies (i.e. seroconvert) within three to twelve weeks of the initial infection. Diagnosis of primary HIV before seroconversion is done by measuring HIV-RNA or p24 antigen. Positive results obtained by antibody or PCR testing are confirmed either by a different antibody or by PCR.

Antibody tests in children younger than 18 months are typically inaccurate due to the continued presence of maternal antibodies. Thus HIV infection can only be diagnosed by PCR testing for HIV RNA or DNA, or via testing for the p24 antigen. Much of the world lacks access to reliable PCR testing and many places simply wait until either symptoms develop or the child is old enough for accurate antibody testing. In sub-Saharan Africa as of 2007–2009 between 30 and 70% of the population were aware of their HIV status. In 2009, between 3.6 and 42% of men and women in Sub-Saharan countries were tested which represented a significant increase compared to previous years.

Classifications

Two main clinical staging systems are used to classify HIV and HIV-related disease for surveillance purposes: the WHO disease staging system for HIV infection and disease, and the CDC classification system for HIV infection. The CDC's classification system is more frequently adopted in developed countries. Since the WHO's staging system does not require laboratory tests, it is suited to the resource-restricted conditions encountered in developing countries, where it can also be used to help guide clinical management. Despite their differences, the two systems allow comparison for statistical purposes.

The World Health Organization first proposed a definition for AIDS in 1986. Since then, the WHO classification has been updated and expanded several times, with the most recent version being published in 2007. The WHO system uses the following categories:

- Primary HIV infection: May be either asymptomatic or associated with acute retroviral syndrome.

- Stage I: HIV infection is asymptomatic with a $CD4^+$ T cell count (also known as CD4 count) greater than 500 per microlitre (μl or cubic mm) of blood. May include generalized lymph node enlargement.

- Stage II: Mild symptoms which may include minor mucocutaneous manifestations and recurrent upper respiratory tract infections. A CD4 count of less than 500/μl.

- Stage III: Advanced symptoms which may include unexplained chronic diarrhea for longer than a month, severe bacterial infections including tuberculosis of the lung, and a CD4 count of less than 350/μl.

- Stage IV or AIDS: severe symptoms which include toxoplasmosis of the brain, candidiasis of the esophagus, trachea, bronchi or lungs and Kaposi's sarcoma. A CD4 count of less than 200/μl.

The United States Center for Disease Control and Prevention also created a classification system for HIV, and updated it in 2008 and 2014. This system classifies HIV infections based on CD4 count and clinical symptoms, and describes the infection in five groups. In those greater than six years of age it is:

- Stage 0: the time between a negative or indeterminate HIV test followed less than 180 days by a positive test

- Stage 1: CD4 count $\geq$ 500 cells/μl and no AIDS defining conditions

- Stage 2: CD4 count 200 to 500 cells/μl and no AIDS defining conditions

- Stage 3: CD4 count $\leq$ 200 cells/μl or AIDS defining conditions

- Unknown: if insufficient information is available to make any of the above classifications

For surveillance purposes, the AIDS diagnosis still stands even if, after treatment, the $CD4^+$ T cell count rises to above 200 per μL of blood or other AIDS-defining illnesses are cured.

Prevention

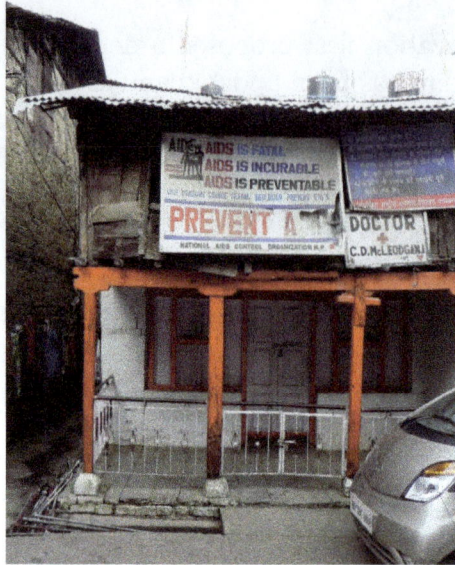

AIDS Clinic, McLeod Ganj, Himachal Pradesh, India, 2010

Sexual Contact

Consistent condom use reduces the risk of HIV transmission by approximately 80% over the long term. When condoms are used consistently by a couple in which one person is infected, the rate of HIV infection is less than 1% per year. There is some evidence to suggest that female condoms may provide an equivalent level of protection. Application of a vaginal gel containing tenofovir (a reverse transcriptase inhibitor) immediately before sex seems to reduce infection rates by approximately 40% among African women. By contrast, use of the spermicide nonoxynol-9 may increase the risk of transmission due to its tendency to cause vaginal and rectal irritation.

Circumcision in Sub-Saharan Africa "reduces the acquisition of HIV by heterosexual men by between 38% and 66% over 24 months". Due to these studies, both the World Health Organization and UNAIDS recommended male circumcision as a method of preventing female-to-male HIV transmission in 2007 in areas with a high rates of HIV. However, whether it protects against male-to-female transmission is disputed, and whether it is of benefit in developed countries and among men who have sex with men is undetermined. The International Antiviral Society, however, does recommend for all sexually active heterosexual males and that it be discussed as an option with men who have sex with men. Some experts fear that a lower perception of vulnerability among circumcised men may cause more sexual risk-taking behavior, thus negating its preventive effects.

Programs encouraging sexual abstinence do not appear to affect subsequent HIV risk. Evidence of any benefit from peer education is equally poor. Comprehensive sexual education provided at school may decrease high risk behavior. A substantial minority of young people continues to engage in high-risk practices despite knowing about HIV/ AIDS, underestimating their own risk of becoming infected with HIV. Voluntary counseling and testing people for HIV does not affect risky behavior in those who test negative but does increase condom use in those who test positive. It is not known whether treating other sexually transmitted infections is effective in preventing HIV.

Pre-exposure

Antiretroviral treatment among people with HIV whose CD4 count ≤ 550 cells/μL is a very effective way to prevent HIV infection of their partner (a strategy known as treatment as prevention, or TASP). TASP is associated with a 10 to 20 fold reduction in transmission risk. Pre-exposure prophylaxis (PrEP) with a daily dose of the medications tenofovir, with or without emtricitabine, is effective in a number of groups including men who have sex with men, couples where one is HIV positive, and young heterosexuals in Africa. It may also be effective in intravenous drug users with a study finding a decrease in risk of 0.7 to 0.4 per 100 person years.

Universal precautions within the health care environment are believed to be effective in decreasing the risk of HIV. Intravenous drug use is an important risk factor and harm reduction strategies such as needle-exchange programs and opioid substitution therapy appear effective in decreasing this risk.

Post-exposure

A course of antiretrovirals administered within 48 to 72 hours after exposure to HIV-positive blood or genital secretions is referred to as post-exposure prophylaxis (PEP). The use of the single agent zidovudine reduces the risk of a HIV infection fivefold following a needle-stick injury. As of 2013, the prevention regimen recommended in the United States consists of three medications—tenofovir, emtricitabine and raltegravir—as this may reduce the risk further.

PEP treatment is recommended after a sexual assault when the perpetrator is known to be HIV positive, but is controversial when their HIV status is unknown. The duration of treatment is usually four weeks and is frequently associated with adverse effects—where zidovudine is used, about 70% of cases result in adverse effects such as nausea (24%), fatigue (22%), emotional distress (13%) and headaches (9%).

Mother-to-child

Programs to prevent the vertical transmission of HIV (from mothers to children) can reduce rates of transmission by 92–99%. This primarily involves the use of a combination of antiviral medications during pregnancy and after birth in the infant and potentially includes bottle feeding rather than breastfeeding. If replacement feeding

is acceptable, feasible, affordable, sustainable, and safe, mothers should avoid breast-feeding their infants; however exclusive breastfeeding is recommended during the first months of life if this is not the case. If exclusive breastfeeding is carried out, the provision of extended antiretroviral prophylaxis to the infant decreases the risk of transmission. In 2015, Cuba became the first country in the world to eradicate mother-to-child transmission of HIV.

Vaccination

Currently, there is no licensed vaccine for HIV or AIDS. The most effective vaccine trial to date, RV 144, was published in 2009 and found a partial reduction in the risk of transmission of roughly 30%, stimulating some hope in the research community of developing a truly effective vaccine. Further trials of the RV 144 vaccine are ongoing.

Treatment

There is currently no cure or effective HIV vaccine. Treatment consists of highly active antiretroviral therapy (HAART) which slows progression of the disease. As of 2010 more than 6.6 million people were taking them in low and middle income countries. Treatment also includes preventive and active treatment of opportunistic infections.

Antiviral Therapy

Current HAART options are combinations (or "cocktails") consisting of at least three medications belonging to at least two types, or "classes," of antiretroviral agents. Initially treatment is typically a non-nucleoside reverse transcriptase inhibitor (NNRTI) plus two nucleoside analogue reverse transcriptase inhibitors (NRTIs). Typical NRTIs include: zidovudine (AZT) or tenofovir (TDF) and lamivudine (3TC) or emtricitabine (FTC). Combinations of agents which include protease inhibitors (PI) are used if the above regimen loses effectiveness.

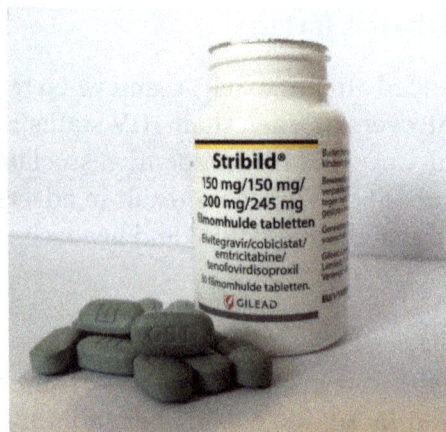

Stribild – a common once-daily ART regime consisting of elvitegravir, emtricitabine, tenofovir and the booster cobicistat

The World Health Organization and United States recommends antiretrovirals in people of all ages including pregnant women as soon as the diagnosis is made regardless of CD4 count. Once treatment is begun it is recommended that it is continued without breaks or "holidays". Many people are diagnosed only after treatment ideally should have begun. The desired outcome of treatment is a long term plasma HIV-RNA count below 50 copies/mL. Levels to determine if treatment is effective are initially recommended after four weeks and once levels fall below 50 copies/mL checks every three to six months are typically adequate. Inadequate control is deemed to be greater than 400 copies/mL. Based on these criteria treatment is effective in more than 95% of people during the first year.

Benefits of treatment include a decreased risk of progression to AIDS and a decreased risk of death. In the developing world treatment also improves physical and mental health. With treatment there is a 70% reduced risk of acquiring tuberculosis. Additional benefits include a decreased risk of transmission of the disease to sexual partners and a decrease in mother-to-child transmission. The effectiveness of treatment depends to a large part on compliance. Reasons for non-adherence include poor access to medical care, inadequate social supports, mental illness and drug abuse. The complexity of treatment regimens (due to pill numbers and dosing frequency) and adverse effects may reduce adherence. Even though cost is an important issue with some medications, 47% of those who needed them were taking them in low and middle income countries as of 2010 and the rate of adherence is similar in low-income and high-income countries.

Specific adverse events are related to the antiretroviral agent taken. Some relatively common adverse events include: lipodystrophy syndrome, dyslipidemia, and diabetes mellitus, especially with protease inhibitors. Other common symptoms include diarrhea, and an increased risk of cardiovascular disease. Newer recommended treatments are associated with fewer adverse effects. Certain medications may be associated with birth defects and therefore may be unsuitable for women hoping to have children.

Treatment recommendations for children are somewhat different from those for adults. The World Health Organisation recommends treating all children less than 5 years of age; children above 5 are treated like adults. The United States guidelines recommend treating all children less than 12 months of age and all those with HIV RNA counts greater than 100,000 copies/mL between one year and five years of age.

Opportunistic Infections

Measures to prevent opportunistic infections are effective in many people with HIV/AIDS. In addition to improving current disease, treatment with antiretrovirals reduces the risk of developing additional opportunistic infections. Adults and adolescents who are living with HIV (even on anti-retroviral therapy) with no evidence of active tuberculosis in settings with high tuberculosis burden should receive isoniazid preventive therapy (IPT), the tuberculin skin test can be used to help decide if IPT is needed. Vaccination against hepatitis A and B is advised for all people at risk of HIV before they become infected; however it may also be given after infection. Trimethoprim/sulfamethoxazole prophylaxis between four and six weeks of age and ceasing breastfeeding in infants born to HIV positive mothers is recommended in resource limited settings. It is also recommended to prevent PCP when a person's CD4 count is below 200 cells/uL and in those who have or have previously had PCP. People with substantial immunosuppression are also advised to receive prophylactic therapy for toxoplasmosis and Cryptococcus meningitis. Appropriate preventive measures have reduced the rate of these infections by 50% between 1992 and 1997.

Diet

The World Health Organization (WHO) has issued recommendations regarding nutrient requirements in HIV/AIDS. A generally healthy diet is promoted. Some evidence has shown a benefit from micronutrient supplements. Evidence for supplementation with selenium is mixed with some tentative evidence of benefit. There is some evidence that vitamin A supplementation in children reduces mortality and improves growth. In Africa in nutritionally compromised pregnant and lactating women a multivitamin supplementation has improved outcomes for both mothers and children. Dietary intake of micronutrients at RDA levels by HIV-infected adults is recommended by the WHO; higher intake of vitamin A, zinc, and iron can produce adverse effects in HIV positive adults, and is not recommended unless there is documented deficiency.

Alternative Medicine

In the US, approximately 60% of people with HIV use various forms of complementary or alternative medicine, even though the effectiveness of most of these therapies has not been established. There is not enough evidence to support the use of herbal medicines. There is insufficient evidence to recommend or support the use of medical cannabis to try to increase appetite or weight gain.

Prognosis

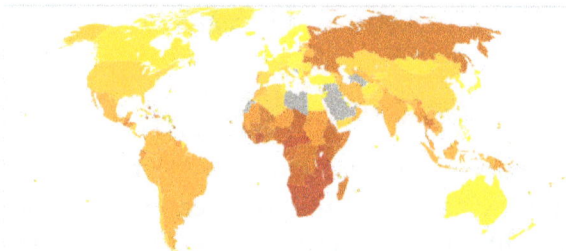

Disability-adjusted life year for HIV and AIDS per 100,000 inhabitants as of 2004.

no data		1000–2500	
≤ 10		2500–5000	
10–25		5000–7500	
25–50		7500-10000	
50–100		10000-50000	
100–500		≥ 50000	
500–1000			

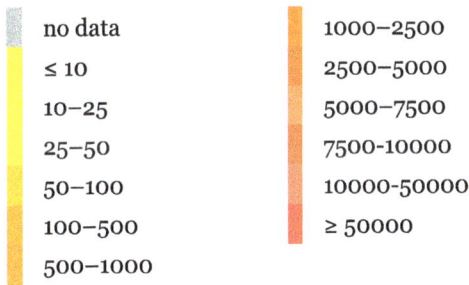

HIV/AIDS has become a chronic rather than an acutely fatal disease in many areas of the world. Prognosis varies between people, and both the CD4 count and viral load are useful for predicted outcomes. Without treatment, average survival time after infection with HIV is estimated to be 9 to 11 years, depending on the HIV subtype. After the diagnosis of AIDS, if treatment is not available, survival ranges between 6 and 19 months. HAART and appropriate prevention of opportunistic infections reduces the death rate by 80%, and raises the life expectancy for a newly diagnosed young adult to 20–50 years. This is between two thirds and nearly that of the general population. If treatment is started late in the infection, prognosis is not as good: for example, if treatment is begun following the diagnosis of AIDS, life expectancy is ~10–40 years. Half of infants born with HIV die before two years of age without treatment.

The primary causes of death from HIV/AIDS are opportunistic infections and cancer, both of which are frequently the result of the progressive failure of the immune system. Risk of cancer appears to increase once the CD4 count is below 500/μL. The rate of clinical disease progression varies widely between individuals and has been shown to be affected by a number of factors such as a person's susceptibility and immune function; their access to health care, the presence of co-infections; and the particular strain (or strains) of the virus involved.

Tuberculosis co-infection is one of the leading causes of sickness and death in those with HIV/AIDS being present in a third of all HIV-infected people and causing 25% of HIV-related deaths. HIV is also one of the most important risk factors for tuberculosis. Hepatitis C is another very common co-infection where each disease increases the progression of the other. The two most common cancers associated with HIV/AIDS are Kaposi's sarcoma and AIDS-related non-Hodgkin's lymphoma.

Even with anti-retroviral treatment, over the long term HIV-infected people may experience neurocognitive disorders, osteoporosis, neuropathy, cancers, nephropathy, and cardiovascular disease. Some conditions like lipodystrophy may be caused both by HIV and its treatment.

Epidemiology

HIV/AIDS is a global pandemic. As of 2014, approximately 37 million people have HIV worldwide with the number of new infections that year being about 2 million. This is down from 3.1 million new infections in 2001. Of these 37 million more than half are women and 2.6 million are less than 15 years old. It resulted in about 1.2 million deaths in 2014, down from a peak of 2.2 million in 2005.

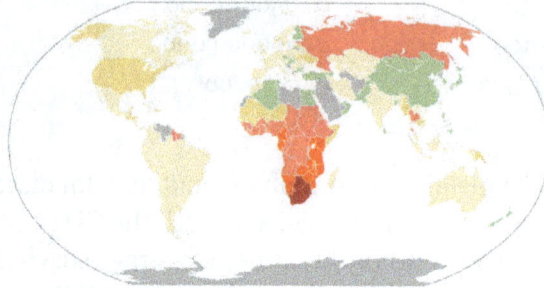

Estimated prevalence in % of HIV among young adults (15–49) per country as of 2011.

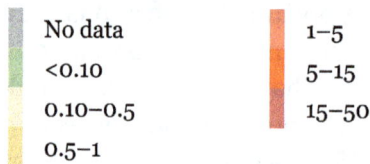

No data		1–5	
<0.10		5–15	
0.10–0.5		15–50	
0.5–1			

Sub-Saharan Africa is the region most affected. In 2010, an estimated 68% (22.9 million) of all HIV cases and 66% of all deaths (1.2 million) occurred in this region. This means that about 5% of the adult population is infected and it is believed to be the cause of 10% of all deaths in children. Here in contrast to other regions women compose nearly 60% of cases. South Africa has the largest population of people with HIV of any country in the world at 5.9 million. Life expectancy has fallen in the worst-affected countries due to HIV/AIDS; for example, in 2006 it was estimated that it had dropped from 65 to 35 years in Botswana. Mother-to-child transmission, as of 2013, in Botswana and South Africa has decreased to less than 5% with improvement in many other African nations due to improved access to antiretroviral therapy.

South & South East Asia is the second most affected; in 2010 this region contained an estimated 4 million cases or 12% of all people living with HIV resulting in approximately 250,000 deaths. Approximately 2.4 million of these cases are in India.

In 2008 in the United States approximately 1.2 million people were living with HIV, resulting in about 17,500 deaths. The US Centers for Disease Control and Prevention estimated that in 2008 20% of infected Americans were unaware of their infection. In the United Kingdom as of 2009 there were approximately 86,500 cases which resulted in 516 deaths. In Canada as of 2008 there were about 65,000 cases causing 53 deaths. Between the first recognition of AIDS in 1981 and 2009 it has led to nearly 30 million deaths. Prevalence is lowest in Middle East and North Africa at 0.1% or less, East Asia at 0.1% and Western and Central Europe at 0.2%. The worst affected European countries, in 2009 and 2012 estimates, are Russia, Ukraine, Latvia, Moldova, Portugal and Belarus, in decreasing order of prevalence.

History

Discovery

The *Morbidity and Mortality Weekly Report* reported in 1981 on what was later to be called "AIDS".

AIDS was first clinically observed in 1981 in the United States. The initial cases were a cluster of injecting drug users and homosexual men with no known cause of impaired immunity who showed symptoms of *Pneumocystis carinii* pneumonia (PCP), a rare opportunistic infection that was known to occur in people with very compromised immune systems. Soon thereafter, an unexpected number of homosexual men developed a previously rare skin cancer called Kaposi's sarcoma (KS). Many more cases of PCP and KS emerged, alerting U.S. Centers for Disease Control and Prevention (CDC) and a CDC task force was formed to monitor the outbreak.

In the early days, the CDC did not have an official name for the disease, often referring to it by way of the diseases that were associated with it, for example, lymphadenopathy, the disease after which the discoverers of HIV originally named the virus. They also used *Kaposi's sarcoma and opportunistic infections*, the name by which a task force had been set up in 1981. At one point, the CDC coined the phrase "the 4H disease", since the syndrome seemed to affect heroin users, homosexuals, hemophiliacs, and Haitians. In the general press, the term "GRID", which stood for gay-related immune deficiency, had been coined. However, after determining that AIDS was not isolated to the gay community, it was realized that the term GRID was misleading and the term AIDS was introduced at a meeting in July 1982. By September 1982 the CDC started referring to the disease as AIDS.

In 1983, two separate research groups led by Robert Gallo and Luc Montagnier declared that a novel retrovirus may have been infecting people with AIDS, and published their findings in the same issue of the journal *Science*. Gallo claimed that a virus his group had isolated from a person with AIDS was strikingly similar in shape to other human T-lymphotropic viruses (HTLVs) his group had been the first to isolate. Gallo's group called their newly isolated virus HTLV-III. At the same time, Montagnier's group isolated a virus from a person presenting with swelling of the lymph nodes of the neck and physical weakness, two characteristic symptoms of AIDS. Contradicting the report from Gallo's group, Montagnier and his colleagues showed that core proteins of this virus were immunologically different from those of HTLV-I. Montagnier's group named their isolated virus lymphadenopathy-associated virus (LAV). As these two viruses turned out to be the same, in 1986, LAV and HTLV-III were renamed HIV.

Origins

Both HIV-1 and HIV-2 are believed to have originated in non-human primates in West-central Africa and were transferred to humans in the early 20th century. HIV-1 appears to have originated in southern Cameroon through the evolution of SIV(cpz), a simian immunodeficiency virus (SIV) that infects wild chimpanzees (HIV-1 descends from the SIVcpz endemic in the chimpanzee subspecies *Pan troglodytes troglodytes*). The closest relative of HIV-2 is SIV(smm), a virus of the sooty mangabey (*Cercocebus atys atys*), an Old World monkey living in coastal West Africa (from southern Senegal to western Côte d'Ivoire). New World monkeys such as the owl monkey are resistant to HIV-1 infection, possibly because of a genomic fusion of two viral resistance genes. HIV-1 is thought to have jumped the species barrier on at least three separate occasions, giving rise to the three groups of the virus, M, N, and O.

There is evidence that humans who participate in bushmeat activities, either as hunters or as bushmeat vendors, commonly acquire SIV. However, SIV is a weak virus which is typically suppressed by the human immune system within weeks of infection. It is thought that several transmissions of the virus from individual to individual in quick succession are necessary to allow it enough time to mutate into HIV. Furthermore, due to its relatively low person-to-person transmission rate, SIV can only spread throughout the population in the presence of one or more high-risk transmission channels, which are thought to have been absent in Africa before the 20th century.

Specific proposed high-risk transmission channels, allowing the virus to adapt to humans and spread throughout the society, depend on the proposed timing of the animal-to-human crossing. Genetic studies of the virus suggest that the most recent common ancestor of the HIV-1 M group dates back to circa 1910. Proponents of this dating link the HIV epidemic with the emergence of colonialism and growth of large colonial African cities, leading to social changes, including a higher degree of sexual promiscuity, the spread of prostitution, and the accompanying high frequency of genital ulcer diseases (such as syphilis) in nascent colonial cities. While transmission rates of HIV during vaginal intercourse are low under regular circumstances, they are increased many fold if one of the partners suffers from a sexually transmitted infection causing genital ulcers. Early 1900s colonial cities were notable due to their high prevalence of prostitution and genital ulcers, to the degree that, as of 1928, as many as 45% of female residents of eastern Kinshasa were thought to have been prostitutes, and, as of 1933, around 15% of all residents of the same city had syphilis.

An alternative view holds that unsafe medical practices in Africa after World War II, such as unsterile reuse of single use syringes during mass vaccination, antibiotic and anti-malaria treatment campaigns, were the initial vector that allowed the virus to adapt to humans and spread.

The earliest well-documented case of HIV in a human dates back to 1959 in the Congo. The earliest retrospectively described case of AIDS is believed to have been in Norway beginning in 1966. In July 1960, in the wake its independence, the United Nations recruited Francophone experts and technicians from all over the world to assist in filling administrative gaps left by Belgium, who did not leave behind an African elite to run the country. By 1962, Haitians made up the second largest group of well-educated experts (out of the 48 national groups recruited), that totaled around 4500 in the country. Dr. Jacques Pépin, a Quebecer author of *The Origins of AIDS*, stipulates that Haiti was one of HIV's entry points to the United States and that one of them may have carried HIV back across the Atlantic in the 1960s. Although, the virus may have been present in the United States as early as 1966, the vast majority of infections occurring outside sub-Saharan Africa (including the U.S.) can be traced back to a single unknown individual who became infected with HIV in Haiti and then brought the infection to the United States some time around 1969. The epidemic then rapidly spread among high-risk groups (initially, sexually promiscuous men who have sex with men). By 1978, the prevalence of HIV-1 among homosexual male residents of New York and San Francisco was estimated at 5%, suggesting that several thousand individuals in the country had been infected.

Society and Culture

Stigma

Ryan White became a poster child for HIV after being expelled from school because he was infected.

AIDS stigma exists around the world in a variety of ways, including ostracism, rejection, discrimination and avoidance of HIV infected people; compulsory HIV testing without prior consent or protection of confidentiality; violence against HIV infected individuals or people who are perceived to be infected with HIV; and the quarantine of HIV infected individuals. Stigma-related violence or the fear of violence prevents many people from seeking HIV testing, returning for their results, or securing treatment, possibly turning what could be a manageable chronic illness into a death sentence and perpetuating the spread of HIV.

AIDS stigma has been further divided into the following three categories:

- *Instrumental AIDS stigma*—a reflection of the fear and apprehension that are likely to be associated with any deadly and transmissible illness.

- *Symbolic AIDS stigma*—the use of HIV/AIDS to express attitudes toward the social groups or lifestyles perceived to be associated with the disease.

- *Courtesy AIDS stigma*—stigmatization of people connected to the issue of HIV/AIDS or HIV-positive people.

Often, AIDS stigma is expressed in conjunction with one or more other stigmas, particularly those associated with homosexuality, bisexuality, promiscuity, prostitution, and intravenous drug use.

In many developed countries, there is an association between AIDS and homosexuality or bisexuality, and this association is correlated with higher levels of sexual prejudice, such as anti-homosexual/bisexual attitudes. There is also a perceived association between AIDS and all male-male sexual behavior, including sex between uninfected men. However, the dominant mode of spread worldwide for HIV remains heterosexual transmission.

In 2003, as part of an overall reform of marriage and population legislation, it became legal for people with AIDS to marry in China.

Economic Impact

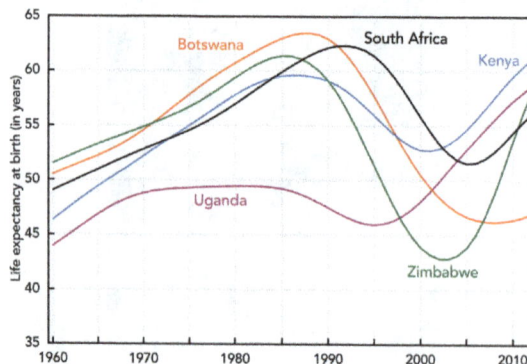

Changes in life expectancy in some African countries, 1960–2012

HIV/AIDS affects the economics of both individuals and countries. The gross domestic product of the most affected countries has decreased due to the lack of human capital. Without proper nutrition, health care and medicine, large numbers of people die from AIDS-related complications. They will not only be unable to work, but will also require significant medical care. It is estimated that as of 2007 there were 12 million AIDS orphans. Many are cared for by elderly grandparents.

Returning to work after beginning treatment for HIV/AIDS is difficult, and affected people often work less than the average worker. Unemployment in people with HIV/AIDS also is associated with suicidal ideation, memory problems, and social isolation; employment increases self-esteem, sense of dignity, confidence, and quality of life. A 2015 Cochrane review found low-quality evidence that antiretroviral treatment helps people with HIV/AIDS work more, and increases the chance that a person with HIV/AIDS will be employed.

By affecting mainly young adults, AIDS reduces the taxable population, in turn reducing the resources available for public expenditures such as education and health services not related to AIDS resulting in increasing pressure for the state's finances and slower growth of the economy. This causes a slower growth of the tax base, an effect that is reinforced if there are growing expenditures on treating the sick, training (to replace sick workers), sick pay and caring for AIDS orphans. This is especially true if the sharp increase in adult mortality shifts the responsibility and blame from the family to the government in caring for these orphans.

At the household level, AIDS causes both loss of income and increased spending on healthcare. A study in Côte d'Ivoire showed that households having a person with HIV/AIDS spent twice as much on medical expenses as other households. This additional expenditure also leaves less income to spend on education and other personal or family investment.

Religion and AIDS

The topic of religion and AIDS has become highly controversial in the past twenty years, primarily because some religious authorities have publicly declared their opposition to the use of condoms. The religious approach to prevent the spread of AIDS according to a report by American health expert Matthew Hanley titled *The Catholic Church and the Global AIDS Crisis* argues that cultural changes are needed including a re-emphasis on fidelity within marriage and sexual abstinence outside of it.

Some religious organisations have claimed that prayer can cure HIV/AIDS. In 2011, the BBC reported that some churches in London were claiming that prayer would cure AIDS, and the Hackney-based Centre for the Study of Sexual Health and HIV reported that several people stopped taking their medication, sometimes on the direct advice of their pastor, leading to a number of deaths. The Synagogue Church Of All Nations advertise an "anointing water" to promote God's healing, although the group deny advising people to stop taking medication.

Media Portrayal

One of the first high-profile cases of AIDS was the American Rock Hudson, a gay actor who had been married and divorced earlier in life, who died on October 2, 1985 having announced that he was suffering from the virus on July 25 that year. He had been diagnosed during 1984. A notable British casualty of AIDS that year was Nicholas Eden, a gay politician and son of the late prime minister Anthony Eden. On November 24, 1991, the virus claimed the life of British rock star Freddie Mercury, lead singer of the band Queen, who died from an AIDS-related illness having only revealed the diagnosis on the previous day. However, he had been diagnosed as HIV positive in 1987. One of the first high-profile heterosexual cases of the virus was Arthur Ashe, the American tennis player. He was diagnosed as HIV positive on August 31, 1988, having contracted the virus from blood transfusions during heart surgery earlier in the 1980s. Further tests within 24 hours of the initial diagnosis revealed that Ashe had AIDS, but he did not tell the public about his diagnosis until April 1992. He died as a result on February 6, 1993 at age 49.

Therese Frare's photograph of gay activist David Kirby, as he lay dying from AIDS while surrounded by family, was taken in April 1990. *LIFE magazine* said the photo became the one image "most powerfully identified with the HIV/AIDS epidemic." The photo was displayed in *LIFE magazine*, was the winner of the World Press Photo, and acquired worldwide notoriety after being used in a United Colors of Benetton advertising campaign in 1992. In 1996, Johnson Aziga, a Ugandan-born Canadian was diagnosed with HIV, but subsequently had unprotected sex with 11 women without disclosing his diagnosis. By 2003 seven had contracted HIV, and two died from complications related to AIDS. Aziga was convicted of first-degree murder and is liable to a life sentence.

Criminal Transmission

Criminal transmission of HIV is the intentional or reckless infection of a person with the human immunodeficiency virus (HIV). Some countries or jurisdictions, including some areas of the United States, have laws that criminalize HIV transmission or exposure. Others may charge the accused under laws enacted before the HIV pandemic.

Misconceptions

There are many misconceptions about HIV and AIDS. Three of the most common are that AIDS can spread through casual contact, that sexual intercourse with a virgin will cure AIDS, and that HIV can infect only gay men and drug users. In 2014, some among the British public wrongly thought you could get HIV from kissing (16%), sharing a glass (5%), spitting (16%), a public toilet seat (4%), and coughing or sneezing (5%). Other misconceptions are that any act of anal intercourse between two uninfected gay men can lead to HIV infection, and that open discussion of HIV and homosexuality in schools will lead to increased rates of AIDS.

A small group of individuals continue to dispute the connection between HIV and AIDS, the existence of HIV itself, or the validity of HIV testing and treatment methods. These claims, known as AIDS denialism, have been examined and rejected by the scientific community. However, they have had a significant political impact, particularly in South Africa, where the government's official embrace of AIDS denialism (1999–2005) was responsible for its ineffective response to that country's AIDS epidemic, and has been blamed for hundreds of thousands of avoidable deaths and HIV infections.

Several discredited conspiracy theories have held that HIV was created by scientists, either inadvertently or deliberately. Operation INFEKTION was a worldwide Soviet active measures operation to spread the claim that the United States had created HIV/AIDS. Surveys show that a significant number of people believed – and continue to believe – in such claims.

Research

HIV/AIDS research includes all medical research which attempts to prevent, treat, or cure HIV/AIDS along with fundamental research about the nature of HIV as an infectious agent and AIDS as the disease caused by HIV.

Many governments and research institutions participate in HIV/AIDS research. This research includes behavioral health interventions such as sex education, and drug development, such as research into microbicides for sexually transmitted diseases, HIV vaccines, and antiretroviral drugs. Other medical research areas include the topics of pre-exposure prophylaxis, post-exposure prophylaxis, and circumcision and HIV.

Swine Influenza

Swine influenza, also called pig influenza, swine flu, hog flu and pig flu, is an infection caused by any one of several types of swine influenza viruses. Swine influenza virus (SIV) or swine-origin influenza virus (S-OIV) is any strain of the influenza family of viruses that is endemic in pigs. As of 2009, the known SIV strains include influenza C and the subtypes of influenza A known as H1N1, H1N2, H2N1, H3N1, H3N2, and H2N3.

Swine influenza virus is common throughout pig populations worldwide. Transmission of the virus from pigs to humans is not common and does not always lead to human flu, often resulting only in the production of antibodies in the blood. If transmission does cause human flu, it is called zoonotic swine flu. People with regular exposure to pigs are at increased risk of swine flu infection.

Around the mid-20th century, identification of influenza subtypes became possible, allowing accurate diagnosis of transmission to humans. Since then, only 50 such transmissions have been confirmed. These strains of swine flu rarely pass from human to human. Symptoms of zoonotic swine flu in humans are similar to those of influenza and of influenza-like illness in general, namely chills, fever, sore throat, muscle pains, severe headache, coughing, weakness and general discomfort.

In August 2010, the World Health Organization declared the swine flu pandemic officially over.

Cases of swine flu have been reported in India, with over 31,156 positive test cases and 1,841 deaths till March 2015.

Signs and Symptoms

Swine

In Swine, an influenza infection produces fever, lethargy, sneezing, coughing, difficulty breathing and decreased appetite. In some cases the infection can cause abortion. Although mortality is usually low (around 1–4%), the virus can produce weight loss and poor growth, causing economic loss to farmers. Infected pigs can lose up to 12 pounds of body weight over a three- to four-week period.

Humans

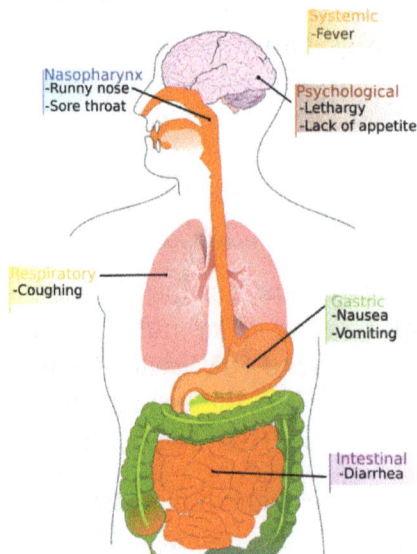

Symptoms of Swine Flu

Systemic
-Fever

Nasopharynx
-Runny nose
-Sore throat

Psychological
-Lethargy
-Lack of appetite

Respiratory
-Coughing

Gastric
-Nausea
-Vomiting

Intestinal
-Diarrhea

Main symptoms of swine flu in humans

Direct transmission of a swine flu virus from pigs to humans is occasionally possible (zoonotic swine flu). In all, 50 cases are known to have occurred since the first report in medical literature in 1958, which have resulted in a total of six deaths. Of these six people, one was pregnant, one had leukemia, one had Hodgkin's lymphoma and two were known to be previously healthy. Despite these apparently low numbers of infections, the true rate of infection may be higher, since most cases only cause a very mild disease, and will probably never be reported or diagnosed.

In this video, Dr. Joe Bresee, with CDC's Influenza Division, describes the symptoms of swine flu and warning signs to look for that indicate the need for urgent medical attention.

According to the Centers for Disease Control and Prevention (CDC), in humans the symptoms of the 2009 "swine flu" H1N1 virus are similar to those of influenza and of influenza-like illness in general. Symptoms include fever; cough, sore throat, watery eyes, body aches, shortness of breath, headache, weight loss, chills, sneezing, runny nose, coughing, dizziness, abdominal pain, lack of appetite and fatigue. The 2009 outbreak has shown an increased percentage of patients reporting diarrhea and vomiting as well. The 2009 H1N1 virus is not zoonotic swine flu, as it is not transmitted from pigs to humans, but from person to person through airborne droplets.

Because these symptoms are not specific to swine flu, a differential diagnosis of *probable* swine flu requires not only symptoms, but also a high likelihood of swine flu due to the person's recent and past medical history. For example, during the 2009 swine flu outbreak in the United States, the CDC advised physicians to "consider swine influenza infection in the differential diagnosis of patients with acute febrile respiratory illness who have either been in contact with persons with confirmed swine flu, or who were in one of the five U.S. states that have reported swine flu cases or in Mexico during the seven days preceding their illness onset." A diagnosis of *confirmed* swine flu requires laboratory testing of a respiratory sample (a simple nose and throat swab).

The most common cause of death is respiratory failure. Other causes of death are pneumonia (leading to sepsis), high fever (leading to neurological problems), dehydration (from excessive vomiting and diarrhea), electrolyte imbalance and kidney failure. Fatalities are more likely in young children and the elderly.

Virology

Transmission

Between Pigs

Influenza is quite common in pigs, with about half of breeding pigs having been exposed to the virus in the US. Antibodies to the virus are also common in pigs in other countries.

The main route of transmission is through direct contact between infected and uninfected animals. These close contacts are particularly common during animal transport. Intensive farming may also increase the risk of transmission, as the pigs are raised in very close proximity to each other. The direct transfer of the virus probably occurs either by pigs touching noses, or through dried mucus. Airborne transmission through the aerosols produced by pigs coughing or sneezing are also an important means of infection. The virus usually spreads quickly through a herd, infecting all the pigs within just a few days. Transmission may also occur through wild animals, such as wild boar, which can spread the disease between farms.

To Humans

People who work with poultry and swine, especially those with intense exposures, are at increased risk of zoonotic infection with influenza virus endemic in these animals, and constitute a population of human hosts in which zoonosis and reassortment can co-occur. Vaccination of these workers against influenza and surveillance for new influenza strains among this population may therefore be an important public health measure. Transmission of influenza from swine to humans who work with swine was documented in a small surveillance study performed in 2004 at the University of Iowa. This study, among others, forms the basis of a recommendation that people whose jobs involve handling poultry and swine be the focus of increased public health surveillance. Other professions at particular risk of infection are veterinarians and meat processing workers, although the risk of infection for both of these groups is lower than that of farm workers.

Interaction with Avian H5N1 in Pigs

Pigs are unusual as they can be infected with influenza strains that usually infect three different species: pigs, birds and humans. This makes pigs a host where influenza viruses might exchange genes, producing new and dangerous strains. Avian influenza virus H3N2 is endemic in pigs in China, and has been detected in pigs in Vietnam, increasing fears of the emergence of new variant strains. H3N2 evolved from H2N2 by antigenic shift. In August 2004, researchers in China found H5N1 in pigs.

These H5N1 infections may be quite common; in a survey of 10 apparently healthy pigs housed near poultry farms in West Java, where avian flu had broken out, five of the pig samples contained the H5N1 virus. The Indonesian government has since found similar results in the same region. Additional tests of 150 pigs outside the area were negative.

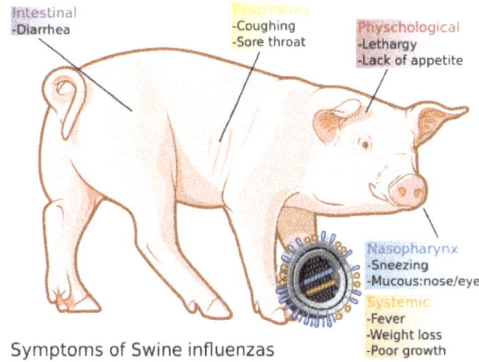

Main symptoms of swine flu in swine

Structure

The influenza virion is roughly spherical. It is an enveloped virus; the outer layer is a lipid membrane which is taken from the host cell in which the virus multiplies. Inserted into the lipid membrane are "spikes", which are proteins—actually glycoproteins, because they consist of protein linked to sugars—known as HA (hemagglutinin) and NA (neuraminidase). These are the proteins that determine the subtype of influenza virus (A/H1N1, for example). The HA and NA are important in the immune response against the virus; antibodies (proteins made to combat infection) against these spikes may protect against infection. The NA protein is the target of the antiviral drugs Relenza and Tamiflu. Also embedded in the lipid membrane is the M2 protein, which is the target of the antiviral adamantanes amantadine and rimantadine.

Classification

Of the three genera of influenza viruses that cause human flu, two also cause influenza in pigs, with influenza A being common in pigs and influenza C being rare. Influenza B has not been reported in pigs. Within influenza A and influenza C, the strains found in pigs and humans are largely distinct, although because of reassortment there have been transfers of genes among strains crossing swine, avian, and human species boundaries.

Influenza C

Influenza viruses infect both humans and pigs, but do not infect birds. Transmission between pigs and humans have occurred in the past. For example, influenza C caused small outbreaks of a mild form of influenza amongst children in Japan and California. Because of its limited host range and the lack of genetic diversity in influenza C, this form of influenza does not cause pandemics in humans.

Influenza A

Swine influenza is caused by influenza A subtypes H1N1, H1N2, H2N3, H3N1, and H3N2. In pigs, four influenza A virus subtypes (H1N1, H1N2,H3N2 and H7N9) are the most common strains worldwide. In the United States, the H1N1 subtype was exclusively prevalent among swine populations before 1998; however, since late August 1998, H3N2 subtypes have been isolated from pigs. As of 2004, H3N2 virus isolates in US swine and turkey stocks were triple reassortants, containing genes from human (HA, NA, and PB1), swine (NS, NP, and M), and avian (PB2 and PA) lineages. In August 2012, the Center for Disease Control and Prevention confirmed 145 human cases (113 in Indiana, 30 in Ohio, one in Hawaii and one in Illinois) of H3N2v since July 2012. The death of a 61-year-old Madison County, Ohio woman is the first in the nation associated with a new swine flu strain. She contracted the illness after having contact with hogs at the Ross County Fair.

Diagnosis

Thermal scanning of passengers arriving at Singapore Changi airport

The CDC recommends real-time PCR as the method of choice for diagnosing H1N1. The oral or nasal fluid collection and RNA virus preserving filter paper card is commercially available. This method allows a specific diagnosis of novel influenza (H1N1) as opposed to seasonal influenza. Near-patient point-of-care tests are in development.

Prevention

Prevention of swine influenza has three components: prevention in swine, prevention of transmission to humans, and prevention of its spread among humans.

Swine

Methods of preventing the spread of influenza among swine include facility management, herd management, and vaccination (ATCvet code: QI09AA03 (WHO)). Because much of the illness and death associated with swine flu involves secondary infection by other pathogens, control strategies that rely on vaccination may be insufficient.

Control of swine influenza by vaccination has become more difficult in recent decades, as the evolution of the virus has resulted in inconsistent responses to traditional vaccines. Standard commercial swine flu vaccines are effective in controlling the infection when the virus strains match enough to have significant cross-protection, and custom (autogenous) vaccines made from the specific viruses isolated are created and used in the more difficult cases. Present vaccination strategies for SIV control and prevention in swine farms typically include the use of one of several bivalent SIV vaccines commercially available in the United States. Of the 97 recent H3N2 isolates examined, only 41 isolates had strong serologic cross-reactions with antiserum to three commercial SIV vaccines. Since the protective ability of influenza vaccines depends primarily on the closeness of the match between the vaccine virus and the epidemic virus, the presence of nonreactive H3N2 SIV variants suggests current commercial vaccines might not effectively protect pigs from infection with a majority of H3N2 viruses. The United States Department of Agriculture researchers say while pig vaccination keeps pigs from getting sick, it does not block infection or shedding of the virus.

Facility management includes using disinfectants and ambient temperature to control viruses in the environment. They are unlikely to survive outside living cells for more than two weeks, except in cold (but above freezing) conditions, and are readily inactivated by disinfectants. Herd management includes not adding pigs carrying influenza to herds that have not been exposed to the virus. The virus survives in healthy carrier pigs for up to three months, and can be recovered from them between outbreaks. Carrier pigs are usually responsible for the introduction of SIV into previously uninfected herds and countries, so new animals should be quarantined. After an outbreak, as immunity in exposed pigs wanes, new outbreaks of the same strain can occur.

Humans

Prevention of pig-to-human transmission

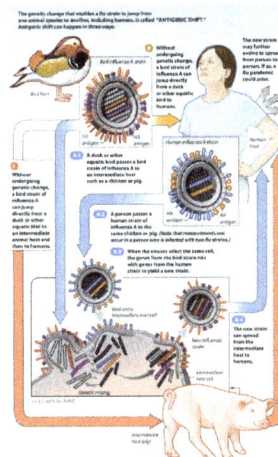

Swine can be infected by both avian and human flu strains of influenza, and therefore are hosts where the antigenic shifts can occur that create new influenza strains.

The transmission from swine to humans is believed to occur mainly in swine farms, where farmers are in close contact with live pigs. Although strains of swine influenza are usually not able to infect humans, this may occasionally happen, so farmers and veterinarians are encouraged to use face masks when dealing with infected animals. The use of vaccines on swine to prevent their infection is a major method of limiting swine-to-human transmission. Risk factors that may contribute to swine-to-human transmission include smoking and, especially, not wearing gloves when working with sick animals, thereby increasing the likelihood of subsequent hand-to-eye, hand-to-nose or hand-to-mouth transmission.

Prevention of human-to-human transmission

Influenza spreads between humans when infected people cough or sneeze, then other people breathe in the virus or touch something with the virus on it and then touch their own face. "Avoid touching your eyes, nose or mouth. Germs spread this way." Swine flu cannot be spread by pork products, since the virus is not transmitted through food. The swine flu in humans is most contagious during the first five days of the illness, although some people, most commonly children, can remain contagious for up to ten days. Diagnosis can be made by sending a specimen, collected during the first five days, for analysis.

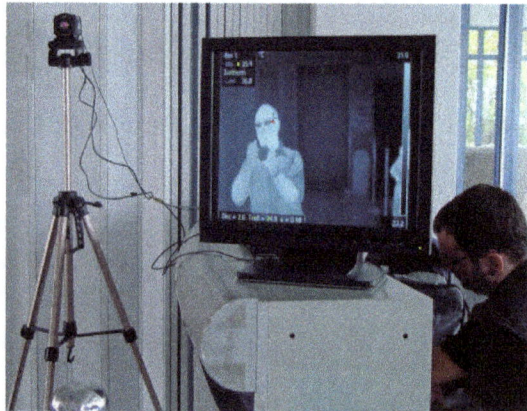

Thermal imaging camera and screen, photographed in an airport terminal in Greece – thermal imaging can detect elevated body temperature, one of the signs of the virus H1N1 (swine influenza).

Recommendations to prevent spread of the virus among humans include using standard infection control, which includes frequent washing of hands with soap and water or with alcohol-based hand sanitizers, especially after being out in public. Chance of transmission is also reduced by disinfecting household surfaces, which can be done effectively with a diluted chlorine bleach solution.

Experts agree hand-washing can help prevent viral infections, including ordinary and the swine flu infections. Also, avoiding touching one's eyes, nose or mouth with one's hands helps to prevent the flu. Influenza can spread in coughs or sneezes, but an increasing body of evidence shows small droplets containing the virus can linger on tabletops, telephones and other surfaces and be transferred via the fingers to the eyes, nose or mouth. Alcohol-based gel or foam hand sanitizers work well to destroy viruses and bacteria. Anyone with flu-like symptoms, such as a sudden fever, cough or muscle aches, should stay away from work or public transportation, and should contact a doctor for advice.

Social distancing, another tactic, is staying away from other people who might be infected, and can include avoiding large gatherings, spreading out a little at work, or perhaps staying home and lying low if an infection is spreading in a community. Public health and other responsible authorities have action plans which may request or require social distancing actions, depending on the severity of the outbreak.

Vaccination

Vaccines are available for different kinds of swine flu. The U.S. Food and Drug Administration (FDA) approved the new swine flu vaccine for use in the United States on September 15, 2009. Studies by the National Institutes of Health show a single dose creates enough antibodies to protect against the virus within about 10 days.

In the aftermath of the 2009 pandemic, several studies were conducted to see who received influenza vaccines. These studies show that whites are much more likely to be vaccinated for seasonal influenza and for the H1N1 strain than African Americans This could be due to several factors. Historically, there has been mistrust of vaccines and of the medical community from African Americans. Many African Americans do not believe vaccines or doctors to be effective. This mistrust stems from the exploitation of the African American communities during studies like the Tuskegee study. Additionally, vaccines are typically administered in clinics, hospitals, or doctor's offices. Many people of lower socioeconomic status are less likely to receive vaccinations because they do not have health insurance.

Surveillance

Although there is no formal national surveillance system in the United States to determine what viruses are circulating in pigs, an informal surveillance network in the United States is part of a world surveillance network.

Treatment

Swine

As swine influenza is rarely fatal to pigs, little treatment beyond rest and supportive care is required. Instead, veterinary efforts are focused on preventing the spread of the virus throughout the farm, or to other farms. Vaccination and animal management techniques are most important in these efforts. Antibiotics are also used to treat this disease, which although they have no effect against the influenza virus, do help prevent bacterial pneumonia and other secondary infections in influenza-weakened herds.

Humans

If a person becomes sick with swine flu, antiviral drugs can make the illness milder and make the patient feel better faster. They may also prevent serious flu complications. For treatment, antiviral drugs work best if started soon after getting sick (within two days of symptoms). Beside antivirals, supportive care at home or in a hospital focuses on controlling fevers, relieving pain and maintaining fluid balance, as well as identifying and treating any secondary infections or other medical problems. The U.S. Centers for Disease Control and Prevention recommends the use of oseltamivir (Tamiflu) or zanamivir (Relenza) for the treatment and/or prevention of infection with swine influenza viruses; however, the majority of people infected with the virus make a full recovery without requiring medical attention or antiviral drugs. The virus isolated in the 2009 outbreak have been found resistant to amantadine and rimantadine.

In the U.S., on April 27, 2009, the FDA issued Emergency Use Authorizations to make available Relenza and Tamiflu antiviral drugs to treat the swine influenza virus in cases for which they are currently unapproved. The agency issued these EUAs to allow treatment of patients younger than the current approval allows and to allow the widespread distribution of the drugs, including by volunteers.

History

Swine influenza was first proposed to be a disease related to human flu during the 1918 flu pandemic, when pigs became ill at the same time as humans. The first identification of an influenza virus as a cause of disease in pigs occurred about ten years later, in 1930. For the following 60 years, swine influenza strains were almost exclusively H1N1. Then, between 1997 and 2002, new strains of three different subtypes and five different genotypes emerged as causes of influenza among pigs in North America. In 1997–1998, H3N2 strains emerged. These strains, which include genes derived by reassortment from human, swine and avian viruses, have become a major cause of swine influenza in North America. Reassortment between H1N1 and H3N2 produced H1N2. In 1999 in Canada, a strain of H4N6 crossed the species barrier from birds to pigs, but was contained on a single farm.

The H1N1 form of swine flu is one of the descendants of the strain that caused the 1918 flu pandemic. As well as persisting in pigs, the descendants of the 1918 virus have also circulated in humans through the 20th century, contributing to the normal seasonal epidemics of influenza. However, direct transmission from pigs to humans is rare, with only 12 recorded cases in the U.S. since 2005. Nevertheless, the retention of influenza strains in pigs after these strains have disappeared from the human population might make pigs a reservoir where influenza viruses could persist, later emerging to reinfect humans once human immunity to these strains has waned.

Swine flu has been reported numerous times as a zoonosis in humans, usually with limited distribution, rarely with a widespread distribution. Outbreaks in swine are common and cause significant economic losses in industry, primarily by causing stunting and extended time to market. For example, this disease costs the British meat industry about £65 million every year.

1918 Pandemic

The 1918 flu pandemic in humans was associated with H1N1 and influenza appearing in pigs; this may reflect a zoonosis either from swine to humans, or from humans to swine. Although it is not certain in which direction the virus was transferred, some evidence suggests, in this case, pigs caught the disease from humans. For instance, swine influenza was only noted as a new disease of pigs in 1918, after the first large outbreaks of influenza amongst people. Although a recent phylogenetic analysis of more recent strains of influenza in humans, birds, animals, and many others and swine suggests the 1918 outbreak in humans followed a reassortment event within a mammal, the exact origin of the 1918 strain remains elusive. It is estimated that anywhere from 50 to 100 million people were killed worldwide.

1976 U.S. Outbreak

On February 5, 1976, a United States army recruit at Fort Dix said he felt tired and weak. He died the next day, and four of his fellow soldiers were later hospitalized. Two weeks after his death, health officials announced the cause of death was a new strain of swine flu. The strain, a variant of H1N1, is known as A/New Jersey/1976 (H1N1). It was detected only from January 19 to February 9 and did not spread beyond Fort Dix.

U.S. President Ford receives a swine flu vaccination

This new strain appeared to be closely related to the strain involved in the 1918 flu pandemic. Moreover, the ensuing increased surveillance uncovered another strain in circulation in the U.S.: A/Victoria/75 (H3N2), which spread simultaneously, also caused illness, and persisted until March. Alarmed public health officials decided action must be taken to head off another major pandemic, and urged President Gerald Ford that every person in the U.S. be vaccinated for the disease.

The vaccination program was plagued by delays and public relations problems. On October 1, 1976, immunizations began, and three senior citizens died soon after receiving their injections. This resulted in a media outcry that linked these deaths to the immunizations, despite the lack of any proof the vaccine was the cause. According to science writer Patrick Di Justo, however, by the time the truth was known—that the deaths were not proven to be related to the vaccine—it was too late. "The government had long feared mass panic about swine flu—now they feared mass panic about the swine flu vaccinations." This became a strong setback to the program.

There were reports of Guillain–Barré syndrome (GBS), a paralyzing neuromuscular disorder, affecting some people who had received swine flu immunizations. Although whether a link exists is still not clear, this syndrome may be a side effect of influenza vaccines. As a result, Di Justo writes, "the public refused to trust a government-operated health program that killed old people and crippled young people." In total, 48,161,019 Americans, or just over 22% of the population, had been immunized by the time the National Influenza Immunization Program was effectively halted on December 16, 1976.

Overall, there were 1098 cases of GBS recorded nationwide by CDC surveillance, 532 of which occurred after vaccination and 543 before vaccination. About one to two cases per 100,000 people of GBS occur every year, whether or not people have been vaccinated. The vaccination program seems to have increased this normal risk of developing GBS by about to one extra case per 100,000 vaccinations.

Recompensation charges were filed for over 4000 cases of severe vaccination damage, including 25 deaths, totalling US\$3.5 billion, by 1979. The CDC stated most studies on modern influenza vaccines have seen no link with GBS, Although one review gives an incidence of about one case per million vaccinations, a large study in China, reported in the *New England Journal of Medicine*, covering close to 100 million doses of H1N1 flu vaccine, found only 11 cases of GBS, which is lower than the normal rate of the disease in China: "The risk-benefit ratio, which is what vaccines and everything in medicine is about, is overwhelmingly in favor of vaccination."

1988 U.S. Outbreak

In September 1988, a swine flu virus killed one woman and infected others. A 32-year-old woman, Barbara Ann Wieners, was eight months pregnant when she and her husband, Ed, became ill after visiting the hog barn at a county fair in Walworth County, Wisconsin. Barbara died eight days later, after developing pneumonia. The only pathogen identified was an H1N1 strain of swine influenza virus. Doctors were able to induce labor and deliver a healthy daughter before she died. Her husband recovered from his symptoms.

Influenza-like illness (ILI) was reportedly widespread among the pigs exhibited at the fair. Of the 25 swine exhibitors aged 9 to 19 at the fair, 19 tested positive for antibodies to SIV, but no serious illnesses were seen. The virus was able to spread between people, since one to three health care personnel who had cared for the pregnant woman developed mild, influenza-like illnesses, and antibody tests suggested they had been infected with swine flu, but there was no community outbreak.

In 1998, swine flu was found in pigs in four U.S. states. Within a year, it had spread through pig populations across the United States. Scientists found this virus had originated in pigs as a recombinant form of flu strains from birds and humans. This outbreak confirmed that pigs can serve as a crucible where novel influenza viruses emerge as a result of the reassortment of genes from different strains. Genetic components of these 1998 triple-hybrid stains would later form six out of the eight viral gene segments in the 2009 flu outbreak.

2007 Philippine Outbreak

On August 20, 2007, Department of Agriculture officers investigated the outbreak of swine flu in Nueva Ecija and central Luzon, Philippines. The mortality rate is less than 10% for swine flu, unless there are complications like hog cholera. On July 27, 2007, the Philippine National Meat Inspection Service (NMIS) raised a hog cholera "red alert" warning over Metro Manila and five regions of Luzon after the disease spread to backyard pig farms in Bulacan and Pampanga, even if these tested negative for the swine flu virus.

2009 Northern Ireland Outbreak

Since November 2009, 14 deaths as a result of swine flu in Northern Ireland have been reported. The majority of the victims were reported to have pre-existing health conditions which had lowered their immunity. This closely corresponds to the 19 patients who had died in the year prior due to swine flu, where 18 of the 19 were determined to have lowered immune systems. Because of this, many mothers who have just given birth are strongly encouraged to get a flu shot because their immune systems are vulnerable. Also, studies have shown that people between the ages of 15 and 44 have the highest rate of infection. Although most people now recover, having any conditions that lower one's immune system increases the risk of having the flu become potentially lethal. In Northern Ireland now, approximately 56% of all people under 65 who are entitled to the vaccine have gotten the shot, and the outbreak is said to be under control.

2015 India Outbreak

Swine flu outbreaks were reported in India in late 2014 and early 2015. As of March 19, 2015 the disease has affected 31,151 people and claimed over 1,841 lives. The largest number of reported cases and deaths due to the disease occurred in the western part of India including states like Delhi, Madhya Pradesh, Rajasthan, and Gujarat. Researchers of MIT have claimed that the swine flu has mutated in India to a more virulent version with changes in Hemagglutinin protein. This has however been disputed by Indian researchers.

2015 Nepal Outbreak

Swine flu outbreaks were reported in Nepal in the spring of 2015. As of April 21, 2015 the disease has claimed 26 lives in the most severely affected district, Jajarkot in North-west Nepal. Cases were also detected in the districts of Kathmandu, Morang, Kaski, and Chitwan. As of 22 April 2015 the Nepal Ministry of Health reported that 2,498 people had been treated in Jajarkot, of whom 552 were believed to have swine flu, and acknowledged that the government's response had been inadequate. The Jajarkot outbreak had just been declared an emergency when the April 2015 Nepal earthquake struck on 25 April 2015, diverting all medical and emergency resources to quake-related rescue and recovery.

2016 Pakistan Outbreak

There were seven cases of Swine flu reported in Punjab province of Pakistan mainly in the city of Multan in January 2016.Cases of Swine Flu have also been reported in Lahore.

H1N1 Virus Pandemic History

A study conducted in 2008, and published in the journal *Nature*, has managed to establish the evolutionary origin of the flu strain of swine origin (S-OIV).

The phylogenetic origin of the flu virus that caused the 2009 pandemics can be traced before 1918. Around 1918, the ancestral virus, of avian origin, crossed the species boundaries and infected humans as human H1N1. The same phenomenon took place soon after in America, where the human virus was infecting pigs; it led to the emergence of the H1N1 swine strain, which later became the classic swine flu.

New events of reassortment were not reported until 1968, when the avian strain H1N1 infected humans again; this time the virus met the strain H2N2, and the reassortment originated the strain H3N2. This strain has remained as a stable flu strain until now.

The mid-1970s were important for the evolution of flu strains. First, the re-emergence of the human H1N1 strain became a seasonal strain. Then, a small outbreak of swine H1N1 occurred in humans, and finally, the human H2N2 strain apparently became extinct. Around 1979, the avian H1N1 strain infected pigs and gave rise to Euroasiatic swine flu and H1N1 Euroasiatic swine virus, which is still being transmitted in swine populations.

The critical moment for the 2009 outbreak was between 1990 and 1993. A triple reassortment event in a pig host of North American H1N1 swine virus, the human H3N2 virus and avian H1N1 virus generated the swine H1N2 strain. Finally, the last step in S-OIV history was in 2009, when the virus H1N2 co-infected a human host at the same time as the Euroasiatic H1N1 swine strain. This led to the emergence of a new human H1N1 strain, which caused the 2009 pandemic.

On June 11, 2009, the World Health Organization raised the worldwide pandemic alert level to Phase 6 for swine flu, which is the highest alert level. This alert level means that the swine flu had spread worldwide and there were cases of people with the virus in most countries. The pandemic level identifies the spread of the disease or virus and not necessarily the severity of the disease.

Swine flu spread very rapidly worldwide due to its high human-to-human transmission rate and due to the frequency of air travel.

In 2015 the instances of Swine Flu substantially increased to five year highs with over 10000 cases reported and 660 deaths in India. The states reporting the highest number of cases and deaths are Rajasthan, Gujarat, Madhya Pradesh, Maharashtra, Delhi, and Telangana. The circulating strain of influenza being the same, unmutant strain that caused global pandemic in 2009 (A H1N1 pdm 09), the sudden spurt of the cases in the beginning of 2015 left the Indian government unexplained but concerned. Government instructed the affected states to investigate into the epidemiological reasons of such spurt in the states, and had detailed the advisory guidelines to all states. The guidelines are mainly for (a) description of A H1N1 for prompt identification, detection, and distinction from the symptoms of other similar infection such as common flu(cold) (b) categorization of screening of influenza A H1N1 cases, (c) clinical management protocol of Pandemic influenza A H1N1, (d) providing home care, (e) collection of human sample. Besides, through the National Centre for Diseases Control (NCDC), Directorate General of Health Services (DGHS), Government of India (GoI) had placed a tender to procure 8 kits of Assay sets, 37 kits of one step RT-PCR kit, and 36 kits of viral RNA extraction kits.

References

- Cheney K, McKnight A (2010). "HIV-2 Tropism and Disease". Lentiviruses and Macrophages: Molecular and Cellular Interactions. Caister Academic Press. ISBN 978-1-904455-60-8.

- Lederberg, editor-in-chief Joshua (2000). Encyclopedia of Microbiology, (4 Volume Set). (2nd ed.). Burlington: Elsevier. p. 106. ISBN 9780080548487. Retrieved 9 June 2016.

- Aldrich, ed. by Robert; Wotherspoon, Garry (2001). Who's who in gay and lesbian history. London: Routledge. p. 154. ISBN 9780415229746.

- Guideline on when to start antiretroviral therapy and on pre-exposure prophylaxis for HIV. (PDF). WHO. 2015. p. 13. ISBN 9789241509565.

- "The impact of AIDS on people and societies" (PDF). 2006 Report on the global AIDS epidemic. UNAIDS. 2006. ISBN 92-9173-479-9. Retrieved June 14, 2006.

- Charles B. Hicks, MD (2001). Jacques W. A. J. Reeders & Philip Charles Goodman, ed. Radiology of AIDS. Berlin [u.a.]: Springer. p. 19. ISBN 978-3-540-66510-6.

- Elliott, Tom (2012). Lecture Notes: Medical Microbiology and Infection. John Wiley & Sons. p. 273. ISBN 978-1-118-37226-5.

- Stürchler, Dieter A. (2006). Exposure a guide to sources of infections. Washington, DC: ASM Press. p. 544. ISBN 978-1-55581-376-5.

- al.], edited by Richard Pattman (2010). Oxford handbook of genitourinary medicine, HIV, and sexual health (2nd ed.). Oxford: Oxford University Press. p. 95. ISBN 978-0-19-957166-6.

- Kerrigan, Deanna (2012). The Global HIV Epidemics among Sex Workers. World Bank Publications. p. 1. ISBN 978-0-8213-9775-6.

- Aral, Sevgi (2013). The New Public Health and STD/HIV Prevention: Personal, Public and Health Systems Approaches. Springer. p. 120. ISBN 978-1-4614-4526-5.

- Martínez, edited by Miguel Angel (2010). RNA interference and viruses : current innovations and future trends. Norfolk: Caister Academic Press. p. 73. ISBN 978-1-904455-56-1.

- Gerald B. Pier, ed. (2004). Immunology, infection, and immunity. Washington, D.C.: ASM Press. p. 550. ISBN 978-1-55581-246-1.

- editor, Julio Aliberti, (2011). Control of Innate and Adaptive Immune Responses During Infectious Diseases. New York, NY: Springer Verlag. p. 145. ISBN 978-1-4614-0483-5.

- Antiretroviral therapy for HIV infection in adults and adolescents: recommendations for a public health approach (PDF). World Health Organization. 2010. pp. 19–20. ISBN 978-92-4-159976-4.

- Consolidated guidelines on the use of antiretroviral drugs for treating and preventing HIV infection (PDF). World Health Organization. 2013. pp. 28–30. ISBN 978-92-4-150572-7.

- Smith, [edited by] Blaine T. (2008). Concepts in immunology and immunotherapeutics (4th ed.). Bethesda, Md.: American Society of Health-System Pharmacists. p. 143. ISBN 978-1-58528-127-5.

- Lederberg, editor-in-chief Joshua (2000). Encyclopedia of Microbiology, (4 Volume Set). (2nd ed.). Burlington: Elsevier. p. 106. ISBN 9780080548487. Retrieved 9 June 2016.

- Jackson, Regine O., ed. (2011). "Geographies of the Haitian Diaspora". Routledge. p. 12. ISBN 9780415887083. Retrieved 13 March 2016.

- "HIV Classification: CDC and WHO Staging Systems". Guide for HIV/AIDS Clinical Care. AIDS Education and Training Center Program. Retrieved November 21, 2015.

- "WHO validates elimination of mother-to-child transmission of HIV and syphilis in Cuba". WHO. June 30, 2015. Retrieved August 30, 2015.

Permissions

All chapters in this book are published with permission under the Creative Commons Attribution Share Alike License or equivalent. Every chapter published in this book has been scrutinized by our experts. Their significance has been extensively debated. The topics covered herein carry significant information for a comprehensive understanding. They may even be implemented as practical applications or may be referred to as a beginning point for further studies.

We would like to thank the editorial team for lending their expertise to make the book truly unique. They have played a crucial role in the development of this book. Without their invaluable contributions this book wouldn't have been possible. They have made vital efforts to compile up to date information on the varied aspects of this subject to make this book a valuable addition to the collection of many professionals and students.

This book was conceptualized with the vision of imparting up-to-date and integrated information in this field. To ensure the same, a matchless editorial board was set up. Every individual on the board went through rigorous rounds of assessment to prove their worth. After which they invested a large part of their time researching and compiling the most relevant data for our readers.

The editorial board has been involved in producing this book since its inception. They have spent rigorous hours researching and exploring the diverse topics which have resulted in the successful publishing of this book. They have passed on their knowledge of decades through this book. To expedite this challenging task, the publisher supported the team at every step. A small team of assistant editors was also appointed to further simplify the editing procedure and attain best results for the readers.

Apart from the editorial board, the designing team has also invested a significant amount of their time in understanding the subject and creating the most relevant covers. They scrutinized every image to scout for the most suitable representation of the subject and create an appropriate cover for the book.

The publishing team has been an ardent support to the editorial, designing and production team. Their endless efforts to recruit the best for this project, has resulted in the accomplishment of this book. They are a veteran in the field of academics and their pool of knowledge is as vast as their experience in printing. Their expertise and guidance has proved useful at every step. Their uncompromising quality standards have made this book an exceptional effort. Their encouragement from time to time has been an inspiration for everyone.

The publisher and the editorial board hope that this book will prove to be a valuable piece of knowledge for students, practitioners and scholars across the globe.

Index

www.ingramcontent.com/pod-product-compliance
Lightning Source LLC
Chambersburg PA
CBHW061949190326
41458CB00009B/2824

* 9 7 8 1 6 3 5 4 9 0 4 7 3 *